STAFFORDSHIRE LIBRARIES, ARTS AND ARCHIVES

KV-013-538

STAFFORDSHIRE
COUNTY REFERENCE
LIBRARY
HANLEY
STOKE-ON-TRENT

WITHDRAWN AND
SOLD BY
STOKE-ON-TRENT
LIBRARIES

OTHER BOOKS BY THE SAME AUTHORS:

Ceramic Painting Step by Step
China Painting Step by Step

Owl Patio Light *by Doris W. Taylor*

Wind Chimes *by Craig Taylor*

Clown Bank *by Jean Taylor*
People Bank *by Sally Swisher*

Dollhouse Food *by Jean Taylor and Sally Swisher*

Candle-holders *by Craig Taylor and Phyllis Swaim*

Bird Feeder *by Doris W. Taylor*

Pixies *by Phyllis Swaim*

Leaf Dishes *by Doris W. Taylor*

CREATIVE CERAMICS
FOR The BEGINNER
STEP BY STEP

Doris W. Taylor

Anne Button Hart

PHOTOGRAPHS BY BOB STAMPER

BURSLEM

D. VAN NOSTRAND COMPANY, INC.
PRINCETON   TORONTO   MELBOURNE   LONDON

VAN NOSTRAND REGIONAL OFFICES: *New York, Chicago, San Francisco*

D. VAN NOSTRAND COMPANY, LTD., *London*

D. VAN NOSTRAND COMPANY, (Canada), LTD., *Toronto*

D. VAN NOSTRAND AUSTRALIA PTY. LTD., *Melbourne*

Copyright © 1968, by AMERICAN BOOK COMPANY

Published simultaneously in Canada by
D. VAN NOSTRAND COMPANY (Canada), LTD.

*No reproduction in any form of this book, in whole
or in part (except for brief quotation in critical articles or
reviews), may be made without written authorization
from the publishers.*

Library of Congress Catalog Card No. 68–9038

PRINTED IN THE UNITED STATES OF AMERICA

PREFACE

Many people are unaware of their creative abilities. Talents can be developed as well as inherited and working with clay is one of the easiest forms of creative expression. What begins as play often will become an absorbing interest and pastime.

Experimenting with clay is both relaxing and stimulating. Also, it is not expensive and requires a minimum of working space. This book is full of step-by-step projects, beginning with the simplest, and your principal tools are YOUR TEN FINGERS!

In order to think creatively with clay you must know something of the technical process. By following the step-by-step instructions with each project you will be able to create clay pieces of interest and beauty.

This book will start you off in the right direction. Then, use your imagination. Have fun. Develop the talent you didn't know you had and don't be surprised if you join the ranks of the professionals someday.

THINK CREATIVELY!

THE AUTHORS

The authors acknowledge with sincere appreciation the assistance of:
> Phyllis Swaim
> Marge Noel
> George Umans

And contributing students:
> Sally Swisher
> Lee Jean Anderson
> Judy Traub
> Craig Taylor
> Ann Hellmuth
> Catherine Hellmuth
> Dora Campbell
> Debbie Hart
> Jean Taylor

The kiln photos and kiln firing schedule are included by courtesy of Paragon Industries, Incorporated

CONTENTS

GENERAL PROCEDURE 1
SUPPLIES 8
FIRING 10
PROJECTS 17
   Pencil Holders 17
      Black Octopus, Square, Clown, People
   Napkin Rings 22
   Trivets, Candy Dishes 25
   Christmas Tree Ornaments 30
   Windchimes 34
      Free Form, Cookie Cutter, Fish, Coil
   Large Leaf Edge Dish 40
   Oak Leaf Ashtray with Acorns 41
   Free Form Dishes and Ashtrays 44
   Birdhouse 49
   Clown Bank 52
   Cat Patio Light 57
   Owl Patio Light 61
   Candle-holder for Patio Light 64
   Bird Feeder 65
   Christmas Figurines 71
      Angel 71
      Three Wise Men 75
   Pansy Ring 84
   Hand Modeling 86
   Horse Head Bookends 88

Dollhouse Food	91
Pixie Dresser Set	95
Grape Candle-holder	101
Holly Candle-holder	104
Rose Candle-holder	106
Coil Items	111
TIPS FOR THE TEACHER	112
COMMON PROBLEMS	114
GLOSSARY	117
INDEX	119

GENERAL PROCEDURE

First, a work area for the projects described step by step on the following pages is relatively easy to set up. Cover the top of a sturdy table with a pillowcase, sheet, newspaper, or an oilcloth bottom side out. Then you won't have to worry about the clay sticking to the tabletop.

There are several types of clay available, but for the projects in this book you must use earthenware clay, which can be fired and glazed. Plastic modeling clay and other types of clay are not suitable. Earthenware clay generally comes in two colors, gray and brick red. Most hobbyists use the gray, which turns white in the kiln, because it doesn't stain the work area—or the worker. The red clay does not significantly change color in firing.

The clay on sale at your local ceramic shop or clay supplier is ready for immediate use. If clay is too wet to handle easily when first removed from the container, roll the clay on a piece of newspaper, which will absorb some of the excess moisture. Turn the clay over and roll on the other side. Repeat until the clay is workable. Dry clay or clay scraps can be conditioned for use by wrapping in a damp terrycloth towel and placing in a plastic bag. The clay will absorb moisture from the towel overnight. Clay with which you are working should be kept in a plastic bag. Slice off pieces as needed with knife. Leave remaining clay in bag or it will dry out.

Clay lumps or clay scraps that have air bubbles need to be wedged. And for this you need a wedging board, which is a relatively easy project for you or the handyman-of-the-house. The three photographs in Figure 1 illustrate the wedging board and the wedging process. To construct a wedging board, fill a flat wooden box approximately eighteen by twenty-four inches and two inches deep with wet plaster of paris. When the plaster has hardened, cover the top of it with a large piece of canvas. This will absorb excess moisture from the clay

  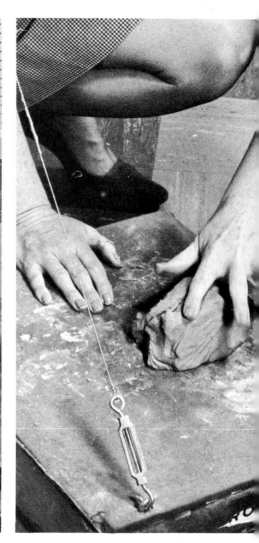

during the wedging process. Then take a piece of standard two-by-four lumber approximately two feet long and nail it upright to the back of the box. Stretch a piece of piano wire (or something similar) from the top of the two-by-four to the front of the box. Secure the wire with a turnbuckle, which can be periodically adjusted to keep the wire taut. Since the wedging board will get some rough use, its construction should be sturdy.

To wedge, take a good-sized lump of clay in your two hands and pass it over the wire so that it is cut in half. Throw first one, then the other half onto the canvas—hard! Repeat the cutting and pounding process a good many times until the clay is even in color and has no air pockets.

If you don't have a wedging board, you might try the following: roll the clay on a newspaper, drop the clay on a newspaper, cut it with wire, and repeat and repeat and repeat—until the clay can be cut without revealing any air pockets. Sometimes, when rolling clay, air pockets will form on the surface. If so, puncture them with a pin and roll again.

Four different methods of handling clay are covered in this book: slab, molding over forms, coil, and hand modeling.

In *slab work,* place the clay between two wooden slats according to chapter instructions. The distances between the slats will vary according to the project. Roll the clay flat. The thickness of the clay also varies with the project. Roll clay in all directions so you don't end up with one narrow strip. If the clay sticks to your fingers, remove some of the moisture by rolling the clay on newspaper. If the clay sticks to your rolling tool (which can be a household rolling pin or piece of dowling from a lumber store), dust the surface of the clay lightly with talcum powder. Remember to wash away all powder

with a damp sponge before the finished piece is put into the kiln, because glaze will not cling over powder.

When you have the clay properly rolled out, trace the paper pattern from this book onto tracing paper or cut a pattern from light cardboard if it is to be used many times. If you trace with an indelible pencil onto tracing paper, place the pattern face down on top of the clay and roll over pattern lightly. The damp clay will absorb pencil marks. Pattern can then be cut out with pointed cutting tool. Don't use a knife. You must have either a "lace tool," available at ceramic shop, or a hat pin.

Often in completing the work on a piece you will have to join one or more pieces of clay together. This can be done by using slip, which serves the purpose of glue. It is simple to make your own slip: take some of the clay being used on the project and add a couple of drops of vinegar (to facilitate drying) and water to the consistency of thick cream.

Rough the clay edges by scoring with pointed tool. There should be a number of ridges or rough spots to which the slip can adhere. Apply the slip, hold the edges firmly together, and smooth over with fingers.

You may wonder about the necessity for such careful joining when you see how easily the two or more pieces of clay seem to adhere to each other as you are working. If you do not carefully score edges and properly apply slip at each and every joint, the finished piece will crack and separate in the kiln.

All pieces must be carefully cleaned with damp sponge before setting aside to dry. Pieces will dry in five to ten days. If you're in a hurry, try two days' natural drying and six to eight hours in a warm oven. See individual project instructions.

Molding over forms is one of the easiest methods of ceramic work for a beginner. You can mold over stones, paper forms, or commercial forms available at your local ceramic store. Roll out clay and drape it over the form. Remember previous instructions on dealing with clay that is too moist or too dry. Completely cover the form with pieces of clay joined with slip. When the form is covered, smooth the clay over with your fingers. Then carefully separate the clay from the form, following the instructions given for each individual project. Using this method, it is easy to make banks, baskets, birdhouses, dishes, vases, and ashtrays. If you model over a paper form and all paper is not easily removed, excess paper will burn away in the kiln without harming the clay.

Coil work is what most of us remember from kindergarten and early grade school. The properly prepared clay is rolled into a cylinder of ever-decreasing diameter until you have something resembling a snake. Then the coils can be formed into such shapes as napkin rings, vases, dishes, and bells of all sizes. Score and use plenty of slip to seal the joints. Of course, the coils may be left exposed, for a rough effect, or smoothed over with your fingers. And when you're finished, clean the entire piece with a damp sponge. Set aside to dry before firing.

Hand modeling is perhaps the most creative type of clay work. You shape the piece into whatever pleases your fancy. Then cut out the inside, or run holes into the center of the object to help in drying and to prevent the piece from exploding in the kiln. When you're finished, smooth with a damp sponge and let dry thoroughly before firing. Obviously, one can make a great array of objects with this method: bookends, pixies, pencil-holders, flowers, candle-holders. Al-

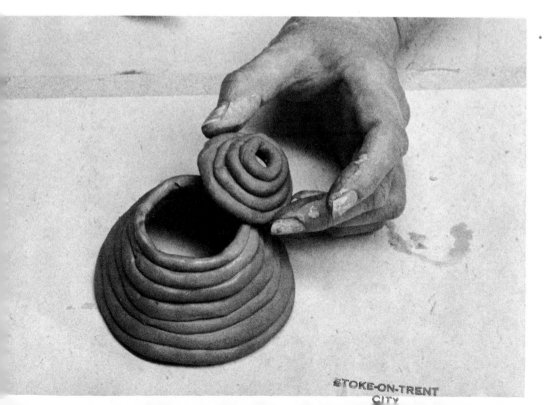

ways leave an escape for trapped air. A hole made with a pointed tool at the top of hat or head is sufficient.

Whatever method you use to create your finished product, the next step is firing. A good many ceramists take their pieces to a commercial kiln—most ceramic stores have one—and leave the firing to them. Other hobbyists prefer having their own kiln for firing.

There are two methods of painting the piece: One, before firing (but after completely dry), with underglaze colors, which must then be fired, painted with clear glaze, and fired again; two, fire, paint with colored glaze, and fire again. Underglaze paints may be purchased in all colors. For strong color they must be applied with brush or sponge in three coats, brushing each coat in a different direction.

You can use either the clear gloss glaze over underglaze paints, which fires to a clear glassy finish, or clear matt glaze, which gives a dull finish after firing. For good coverage, glazes must be applied with brush or sponge in three thin coats, each coat brushed in a different direction. Be careful in applying matt glaze to keep it from piling up in one spot and from running. These imperfections, along with obvious brush marks, will be emphasized in the firing, and the finished piece will be marred.

With underglaze colors you can achieve a variety of effects. Beginners who want to use more than one color on a piece should use underglaze. There is less chance for error than with colored glazes.

After underglazing, fire to 05 cone and then coat with clear glaze. As you advance and wish to use two or more colored glazes, keep the edges thin where the glazes meet. Apply only one coat on the first $1/8$ inch area where glazes meet, but make sure glazes do meet or a white line of separation will show after firing.

It's sometimes better when working with a delicate piece of greenware (unfinished, unfired pottery) to dry and fire it to 05 cone before applying underglaze.

Most of the projects in this book use one color; therefore, colored glazes are specified for the most part. Glazes come in all colors, and should be applied according to the instructions on the jar. Most glazes fire to 06 cone unless otherwise specified. Some glazes come with crystals or you can add your own crystals to the glaze to impart interesting textures to the finished piece. If you use crystals, don't let them pile up at the bottom of a piece. Keep them loose toward the top, since crystals melt and run in the firing process. Speckled and running glazes are also available.

Both glaze and underglaze can be thinned with water. Always wipe the edge of the jars clean before replacing the lid. Wash your brushes carefully in water, but don't let them stand in water too long—they'll lose their shape.

FIRING

Any clay piece, if not completely dry, can explode in the kiln. Fire a solidly molded piece separately after drying four to six weeks, and be sure to dry all other clay pieces a week before firing. Drying can be hurried by placing the piece in a warm oven for six to eight hours after two days' normal drying, or by placing the piece on top of the hot kiln.

Pieces coated with glaze must be raised on metal-pointed stilts to prevent the glaze from sticking to the bottom of the kiln. The small pieces with holes in them, such as the windchimes and Christmas ornaments, must be strung on nichrome wire. Any other wire will melt. The wire must be suspended between posts so that no glazed surface touches another during firing. You should "dryfoot" large pieces by sanding the glaze from the bottom of the piece. Refer to Firing Chapter for further instructions. (See illustration, page 33.)

SUPPLIES

Some projects require additional supplies other than the ones commonly used. These will be listed at the beginning of the specific project. Your local ceramic shops (see the yellow pages of your telephone book) have all the necessary equipment and materials; any one of these three magazines offers additional information on supplies:

Ceramic Arts & Crafts
11408 Greenfield
Detroit, Michigan 48227

Ceramics Monthly
4175 North High Street
Columbus, Ohio 43214

Popular Ceramics Magazine
6011 Santa Monica Boulevard
Los Angeles, California 90038

YOU WILL NEED

Earthenware clay, available from the local ceramic shop.

Rolling tool. Rolling pin or dowling from lumber store. A hardwood shovel handle, which is available at the hardware or lumber store, can be cut into three rolling tools.

$\frac{1}{4}$ inch lattice pieces, 10 inches long, two pieces to a set. These are used for rolling clay to proper thickness. A yardstick cut in two is suitable. Lattice pieces and yardsticks may be purchased at hardware or lumber store. Each person should have two sets, since some clay items are cut double thick ($\frac{1}{2}$ inch).

Old sheet or pillowcase to protect work table and prevent clay from sticking. A large sheet to cover full table is ideal.

Smock to protect clothes.

Terrycloths or paper towels to wipe brushes.

Large box for storing supplies.

Cutting tool—a pointed lace tool available from ceramic shop; a hat pin stuck through a cork will also serve the purpose.

Small natural sponges.

Nylon net (half yard).

Screen—small piece of wire window screen will do.

Sgraffito tool or small kitchen knife for cleaning rough edges.

Underglaze colors.

Clear gloss glaze, clear matt glaze, colored glazes.

Onionskin tracing paper or wax paper.

Lead pencil and/or indelible pencil.

1½ inch brush for glaze.

#12 round brush for underglaze.

#2 round brush for underglaze. A variety of small brushes is desirable so you can use one brush for each color and avoid continually washing brushes.

#0 pointer for trim when applying underglaze.

Patterns. You may trace directly from this text. Or use coloring books, greeting cards, etc.

Vinegar for making slip.

Talcum powder—a small can of baby powder is ideal.

Pencils and ballpoint pens for making holes.

Ceiling tile or small pieces of plywood from lumberyard on which to build projects and for storing them while drying.

Sandpaper. Make a board 12 inches square covered with sandpaper for dryfooting pieces. (See page 14.)

Kiln. This is optional equipment. See chapter on firing.

Carborundum stone—for smoothing bottom of glazed ware when stilts stick during firing—available at hardware store as a knife sharpener.

FIRING

As we noted in the introductory chapter, it's generally easy enough to have your pieces fired by your local ceramic shop or hobby center. But having your own kiln offers many advantages. We recommend that the minimum size be 13 inches wide and 2 feet deep. Buy the largest kiln you can afford; you'll find it worthwhile in the long run. One caution: heavy-duty wiring is required for electric kilns. Check your power availability before making any purchase.

It is desirable to purchase a kiln with an automatic shutoff. Otherwise you'll have to give almost constant attention to the kiln during the firing period. You can't time a piece of pottery as you can a roast—two hours and it's done to a turn. It all depends on the temperature inside the kiln. Ceramists use pyrometric cones to indicate heat intensity. These cones are numbered in series from 022 through 05, and soften and bend at different temperatures. Placing the properly numbered cone inside the kiln at firing time is a necessity if your pieces are going to come out as they should. With an automatic shutoff attachment, the kiln power is cut off at the moment the pyrometric cone softens and bends to a certain angle. Without the automatic shutoff the ceramist must keep looking through the peephole (all kilns have them) to see if the cone is bent. The shutoff is usually dependable, but an added time check is wise to be sure it has completed the job. Without the automatic shutoff, use three cones for guides—one that will soften just before the proper temperature, one at the proper temperature, and one just after.

The kiln is not a household appliance out of the magazine ads; it is a working piece of machinery. Temperatures required for the proper firing of ceramics are well above the melting point of most metals, which means that tremendous stresses are generated during firing. Expansion and contraction of the insulating firebrick will cause cracks at the joints in the brick when the kiln is cold. These cracks will seal as the brick expands during heating.

A properly cared for kiln will give many years of satisfactory service. Maintain the kiln as indicated in the manufacturer's instruction manual. When you do not have an automatic shutoff, place your cones in cone holders and keep them visible from the peephole. Cones should be tilted at an angle of 8–12 degrees from the vertical.

Once you have your kiln, it's important to put it in the best possible location. Remember that a kiln gets quite hot during firing. Keep it at least a foot away from any wall and on a surface that will not be damaged by heat. Avoid rubber tiles or linoleum. Do not have any combustible material near the kiln. *Safety first!*

Vibration in shipping may sometimes cause the elements in the kiln to become dislodged. Replace the elements in the grooves by running a kitchen knife completely around each groove before the kiln is fired. Before using the new kiln, clean it throughly with a damp sponge or the nozzle attachment of a vacuum cleaner. Then protect the bottom of the kiln and the tops of its shelves from glaze dripping with kiln wash, applied with an ordinary paint brush. Keep the wash away from the sides and top of the kiln.

It should be noted that the kiln's shelves, posts, and stilts are made of fireclay. All these items have been fired to a higher temperature than the kiln will reach. Like pottery, they can be broken, so handle with care. Using shelves supported by posts you can materially increase the capacity of your kiln. Half shelves will enable one tall piece to be fired beside two layers of short pieces. Shelves come in many different sizes and shapes. Posts are sold in lengths of even inches. Never use triangular posts taller than ten inches or square posts longer than 14 inches. Remember: the shorter the post, the greater the stability.

Glazed pieces in the kiln are rested on stilts, which are small clay supports with heat-resisting-metal points on their upper surface. Stilts come in a variety of shapes and sizes. After firing, some stilts may be stuck by a thin film of glaze to the fired piece. Carefully break the stilt loose and grind any rough edge smooth with Carborundum stone.

Almost inevitably, there will be some moisture present in your new kiln, so it's a good idea to preheat. Close the kiln and turn the heat switches to low. Leave out the peephole plugs for a moisture vent and heat for about four hours. A cone is not required for this preheating, but you should check the kiln every hour to make sure

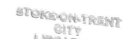

that there is no electric malfunction. Allow the kiln to cool overnight with the lid closed.

The first firing after the preheating should be to cone 018, with the kiln empty except for shelves and cone. The second firing should be to cone 05.

During the first few weeks of the operation of your kiln, check and note inside color frequently so you can estimate inside temperature and get the look and feel of its operation. All manufacturers include complete instructions with their kilns and you should see that your kiln is performing in accordance with these guidelines. Never leave your kiln unattended until you have fired it many times and know its pattern.

Only one pyrometric cone is needed with a kiln equipped with an automatic shutoff device. Nevertheless, it is wise during the first few firings to place additional cones in different areas of the kiln to check on the heat distribution. (It will be obvious which cones were in the area of greatest heat by noting their degree of bend after the kiln has been shut off and opened.) One cone should go on the bottom of the kiln, 3 inches below the peephole in an area where it will be visible. A short post or scrap of firebrick can be used to raise the cone to the proper height. The second cone should be placed on a shelf so as to be visible through the top peephole. Be sure that the cones are 2–4 inches away from the peephole to keep them away from the outside air.

Be sure that all pieces are thoroughly dry before firing. Each project in this book has specific drying instructions. No piece should be placed closer than 1 inch to any heating coil. Larger pieces that need the full kiln dimension should be placed so that the projecting edges are between coils. Leave enough space around pieces to allow for air circulation.

There are various firing techniques used with the different types of ware. Ceramic greenware is the fully formed clay object that has not been fired. See the projects for specific cones used in firing greenware. Plan the load of greenware for the kiln so that the short pieces are on the floor and pieces of similar height are on the same shelves. Greenware can be stacked so that pieces touch each other, but do not stack flat items on top of each other. Pieces should be placed in the positions they will be used in when finished (but wall plaques and the like should be fired with the vertical surface flat to prevent warping).

Apply Kiln Wash to shelves and bottom.

Use posts to position pyrometric cone.

Venting kiln with or without hinged lid. Note: Post does not touch brick.

Typical ceramic bisque load.

Typical ceramic glaze load.

Remove sharp points of glaze left by stilts by rubbing with Carborundum stone.

Ceramic greenware should never be fired with pieces that have been glazed unless they fire at the same cone.

Loading pieces that have been glazed is a tricky business. First, be careful that no piece touches another or anything else to which it might adhere during the firing. Use stilts to support each glazed piece. "Dryfooting" is a technique whereby all glaze is removed from the bottoms of objects with fine sandpaper. This is useful when dealing with items like birdhouses or patio lights that can't be effectively stilted. Be sure to follow the drying instructions for glaze. And since dust on the glazed ware will cause imperfections called pinholes, keep the underside of the kiln shelves and all kiln surfaces clean. This can be done with the nozzle of the vacuum cleaner.

Some decoration is applied over the fired glaze such as gold, lusters, china paints. These require less heat, 018 cone. Place pieces for overglaze firing into the kiln so that no surfaces touch. Use stilts if necessary. Plates should be supported by a rack or placed on edge to permit even firing. Tall posts can separate plates from the kiln walls.

DO'S AND DONT'S ABOUT FIRING THE KILN

Do be sure to fire bisque to 05 to eliminate any chance of the fired piece absorbing moisture. Moisture absorbed after applying glaze will cause the piece to expand and crack. This is called crazing. Fire bisque one cone hotter than glazed ware or even more if glaze can still be applied easily. Slow firing of heavy pieces reduces danger of breakage and crazing. Be sure cone is clearly visible during firing. Do not insert peephole plugs in kiln until switch is on HIGH.
Do make sure the cone is visible during firing.
Do shield your eyes with dark sunglasses if glare makes it difficult to see the cone.
Do turn off the kiln immediately if you can't see the cone.
Do check fuses and/or circuit breakers if electric kiln stops firing.
Do use new cone if ware must be refired due to power failure.

Don't fire by time or pyrometer. Always use cones.
Don't remove ware from kiln until it is cool enough to handle.
Don't allow glaze or kiln wash to touch the electric elements.

On the following page is a chart of times for firing different types of pieces, with switch positions and cone numbers.

WARNING! Never fire by time alone, use a pyrometric cone. If you cannot see the cone, cut off the kiln at once!		START	1 HOUR	1 1/2 HRS	2 HOURS	2 1/2 HRS	3 HOURS	4 HOURS	5 HOURS	6 HOURS
CHINA PAINTS & GOLD	EXTENSION SWITCH	OFF	LOW	LOW	LOW	HIGH				
	TOP SWITCH	LOW	LOW	MEDIUM	MEDIUM	HIGH				
	BOTTOM SWITCH	LOW	MEDIUM	MEDIUM	HIGH	HIGH				
	PEEPHOLE PLUGS	OUT	OUT	OUT	IN*	IN*				
	LID POSITION	VENTED	VENTED	VENTED	CLOSED*	CLOSED				
CERAMIC BISQUE & GLAZE	EXTENSION SWITCH	OFF	OFF	LOW	LOW	LOW	LOW	HIGH		
	TOP SWITCH	LOW	LOW	LOW	MEDIUM	MEDIUM	MEDIUM	HIGH		
	BOTTOM SWITCH	LOW	MEDIUM	MEDIUM	MEDIUM	MEDIUM	HIGH	HIGH		
	PEEPHOLE PLUGS	OUT	OUT	OUT	OUT	OUT	IN*	IN		
	LID POSITION	VENTED	VENTED	VENTED	CLOSED*	CLOSED	CLOSED	CLOSED		

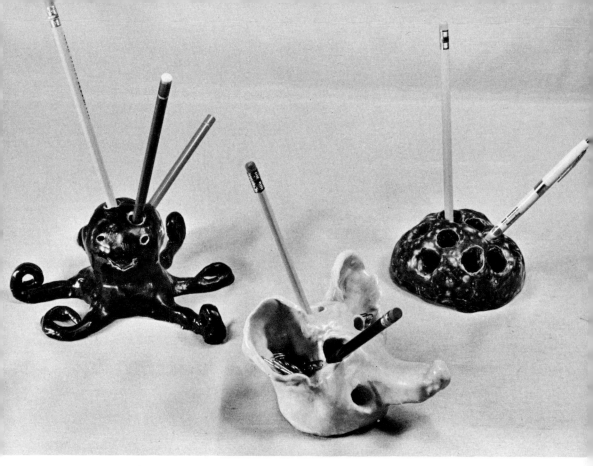

Three kinds of pencil-holders

Black Octopus Pencil-holder
(or Paperweight)

Claywork 30 minutes
Dry 5–7 days

Fire 05 cone
Glaze
Fire 06 cone

MATERIALS

clay the size of grapefruit
pointed tool

ceiling tile
¾ inch brush
4 wooden slats, 10 inches long

black glaze
rolling tool

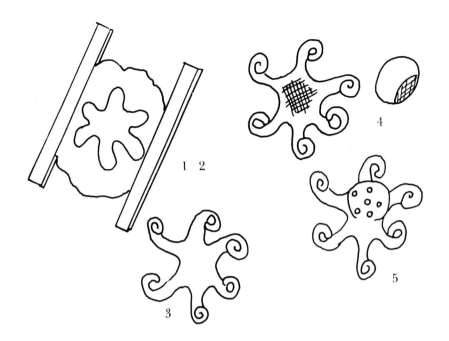

PROCEDURE

STEP 1   Place wooden slats 8 inches apart, two deep for ½ inch thickness. Roll clay between slats.

STEP 2   Trace base from pattern A onto wax or tracing paper. Cut paper pattern with scissors, then place paper pattern on clay and cut with pointed tool.

STEP 3   Pull out legs with fingers and curl up legs 1, 3, and 5. Bend legs 2, 4, and 6 around flat as shown.

STEP 4   Make a clay ball the size of small apple. Flatten one side by pressing it down on the table. Using pointed tool, score bottom side of ball where legs are to be attached. Score top sides of legs where ball is to be attached. Coat scored areas with slip. Place ball on top of legs and smooth joints with fingers.

STEP 5   Using pencil, push in holes for pencils. Holes must go through to bottom or octopus will not fire properly. Dry five to seven days, or after two days' normal drying, place in warm oven six to eight hours.

   Fire to 05 cone.

STEP 6   Using ¾ inch brush, glaze with black glaze and fire to 06 cone. Be sure that glaze coats inside holes.

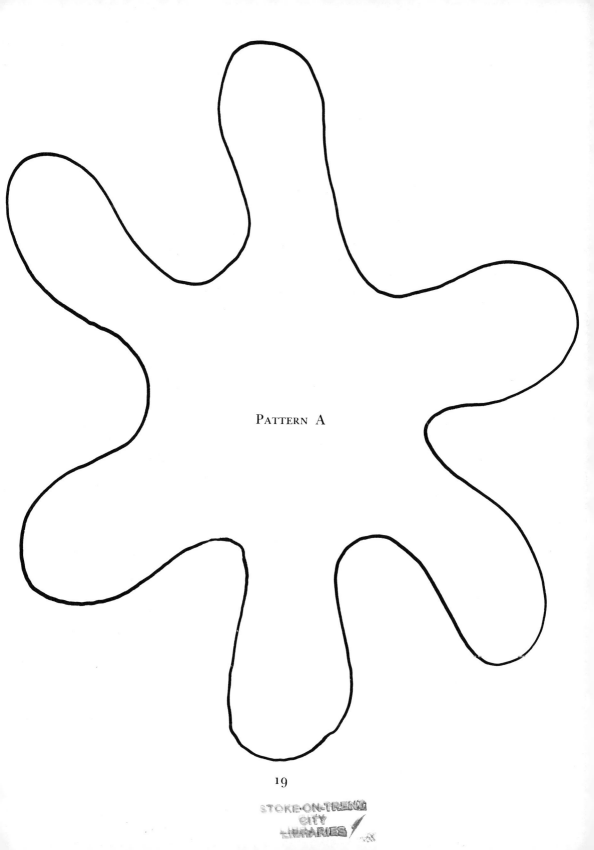

Alternate Suggestions

A SQUARE PENCIL-HOLDER

Take a ball-sized piece of clay and make square on all sides. Poke full of holes from top side through to bottom. Dry on board or ceiling tile for a week, then fire to 05 cone. Glaze with colored glaze, using 3/4 inch brush. Fire to 06 cone.

A CLOWN PENCIL-HOLDER

Refer to section on Banks for general instructions. Make clown solid instead of with a paper core, and then poke in pencil holes. Dry on board for a week; then paint with underglaze colors and fire to 05 cone. Glaze with clear glaze, using 3/4 inch brush. Fire to 06 cone.

A PEOPLE PENCIL-HOLDER

Refer to section on Banks for drawing, but make clay ball solid. Shape mouth, nose, eyes, ears, and feet and make pencil holes. Dry on ceiling tile for a week; paint with underglaze colors; fire to 05 cone. Glaze with clear glaze, using 3/4 inch brush. Fire to 06 cone.

Pencil-holders

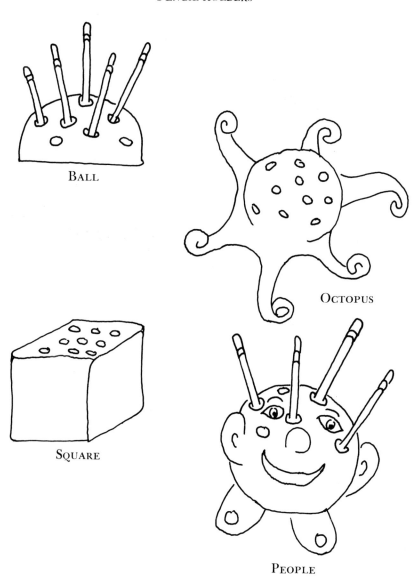

Ball

Octopus

Square

People

Napkin Rings

Claywork: 30 minutes
Dry 5–7 days
Paint with underglaze colors

Fire 05 cone
Glaze with clear gloss or matt
Fire 06 cone

MATERIALS

clay
rolling tool
2 wooden slats
   10 inches long

sponge
slip
screen
cutting tool

underglaze paints
clear glaze
brushes

PROCEDURE

STEP 1　Roll out clay between wooden slats and cut out pattern A. Cut as many shapes as possible from the piece rolled. Shape each strip in the form of ring. Score ends, coat with slip, and seal. Smooth edges and seam with damp sponge.

STEP 2　Press clay scraps to $1/8$ inch thickness and cut butterfly from pattern B. Score back, coat with slip, and attach to ring. Smooth edges with fingers and damp sponge.

STEP 3　Cut three flower pieces from pattern C. Smooth back of flower pieces with damp sponge. Score backs of pieces, coat with slip, and attach to ring. Force a small piece of clay through a screen or a nylon net to make flower centers. Score and coat with slip to attach flower centers.

STEP 4　Carefully clean all edges with damp sponge. Put aside to dry five to seven days, or two days' natural drying and a warm oven six to eight hours.

STEP 5　Paint with underglaze.

STEP 6　Fire 05 cone.

STEP 7　Apply three thin coats of clear glaze.

STEP 8　Fire 06 cone.

You may trim the napkin rings with roses from pp. 106–110, or you may scratch on names by using a sgraffito tool after the ring is leather-hard.

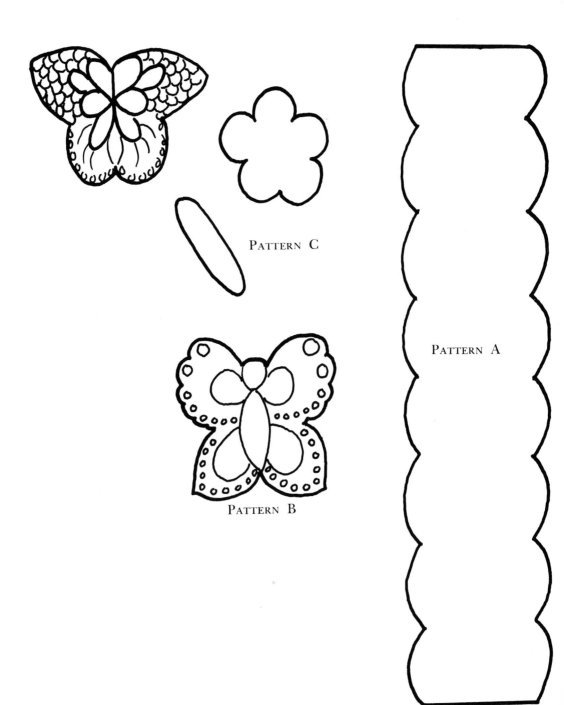

Trivets and Candy Dishes from Rough Surface Patterns

Claywork 15 minutes for Trivet
Dry 5–7 days

Fire 05 cone
Glaze
Fire 06 cone

MATERIALS

plastic or paper doilies
flat metal trivets
clay

4 wooden slats 10 inches long
cutting tool
rolling tool
ceiling tile

talcum
underglaze paints and clear glaze
or colored glaze
brushes

PROCEDURE FOR MAKING A CLAY TRIVET FROM A METAL TRIVET

STEP 1 Take clay the size of small apple. Place slats two deep, six inches apart, and roll clay between. Clay should be ½ inch thick.

STEP 2 To press the design, powder clay lightly, then press trivet, top side down, firmly into clay so that clay squeezes up 1/16 inch through holes in trivet. Powder will keep clay from sticking.

STEP 3 With cutting tool, carefully cut around trivet.

STEP 4 Lift trivet. Wash edges of clay with wet sponge. Be sure to wipe underside of clay trivet smooth with damp sponge; place on ceiling tile to dry. When clay is dry enough so design will not distort, place heavy tile or brick on top to prevent warping while drying.

Wash excess powder from clay trivet carefully with damp sponge.

STEP 5 Dry five to seven days, or two days' natural drying and a warm oven six to eight hours.

STEP 6 Fire 05 cone.

STEP 7 Glaze with green-speckled or any variegated glaze.

STEP 8 Fire 06 cone.

PROCEDURE FOR MAKING DISHES FROM DOILIES

Claywork 2 hours
Dry 5–7 days
Fire 05 cone

Glaze
Fire 06 cone

MATERIALS

plastic or paper doilies
cutting tool
2 wooden slats 10 inches long
clay
sponge
rolling tool
colored underglazes
clear glaze, matt or gloss
brushes
bowl to dry dish in

STEP 1 Roll out clay between slats to a size depending on the size of doily. Place doily on clay, smooth side up. Remove slats. Roll doily firmly into clay.

STEP 2 Carefully cut around doily. Let dry one hour before applying underglaze.

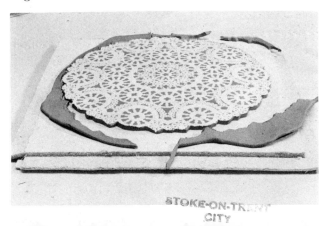

STEP 3 Apply heavy coat of underglaze over doily.

STEP 4 Let dry fifteen to twenty minutes, then remove doily. If paper doily has been used, carefully peel up from edge and tear in toward center. The doily will be destroyed. If available, plastic doilies are easier to work with and may be washed and reused.

STEP 5 Clean edge with damp sponge.

STEP 6 Put into shallow bowl with flat bottom, or invert a smaller dish inside bowl so as to give a flat bottom surface. Finished bowl must have a flat area on the bottom so it will not rock.

STEP 7 Dry five to seven days, or two days' natural drying and a warm oven six to eight hours.
STEP 8 Fire 05 cone.
STEP 9 Glaze with clear matt or clear gloss glaze.
STEP 10 Fire 06 cone.

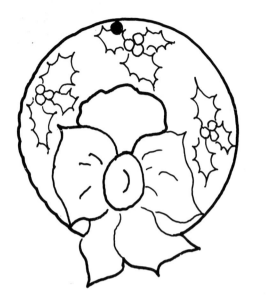

Christmas Ornaments Using Cookie Cutters

Claywork 30 minutes
Dry overnight, before painting
Underglaze

Dry 3–4 days
Fire 05 cone
Glaze, clear or matt
Fire 06 cone

MATERIALS

clay
2 wooden slats
  10 inches long
rolling tool
sponge
tracing paper or
  wax paper
towel
cutting tool
underglaze paints
indelible pencil
pencils or ballpoint
  pen
ceiling tile
talcum powder
brushes
glaze
cookie cutters

metal cookie cutters, open at top and as deep as possible. (If the clay sticks and cutter is open at top, you can push clay loose. If you have flat plastic or metal cutters, use them only to mark the clay, and then cut with cutting tool. Small children do better with simple cutters such as stars, circles, and bells.)

PROCEDURE

STEP 1 Place cloth on worktable and place clay between wooden slats 12 inches apart. Powder the roller slightly and also the clay. Flatten clay with roller and roll clay up and down until it is the same thickness as the slats.

STEP 2 Dip cutters in powder before placing on clay to prevent sticking. Press into clay lightly. *Do not press too hard* or clay will stick to cutter. Remove cutter. Cut design free with cutting tool. With deep metal cutters you may cut down through clay and shake clay loose from cutter. If you use the designs shown, you may trace with indelible pencil. Turn tracing over on damp clay and the design will transfer onto clay when pressed gently.

STEP 3 Make hole in top of ornament with top of ballpoint pen. For trimmings, use cutting tool to score shallow cuts where decoration is to be added. Coat scoring with slip and apply trimming. Slip will make the decorations adhere.

STEP 3 *cont'd* Make little balls of clay and trim, using slip each time. Roll out strings of clay and trim, using slip each time. Scratch freehand designs on ornaments with a pen point.

STEP 4 With damp sponge smooth ornament and remove any excess powder. Place on ceiling tile to dry. Place another weighted tile on top to prevent warping.

When dry, use cutting tool and damp sponge to smooth rough edges. Make sure hole is open in top of ornament.

STEP 5 Dry overnight or longer before painting with underglaze paints, or to make ornaments less breakable, fire to 06 cone before painting.

STEP 6 Apply underglaze paints.
Dry three to five days.

STEP 7 Fire to 05 cone.

STEP 8 Coat with three thin coats of clear glaze.

STEP 9 Fire to 06 cone. When firing glazed pieces, use nichrome wire. Hang pieces on wire, supported by kiln poles. Pieces must not touch during firing.

It's easy to glaze by holding ornament with pencil point in holes.
Cookie-cutter ornaments will also make nice shade pulls. Use above instructions and paint with colored glazes.

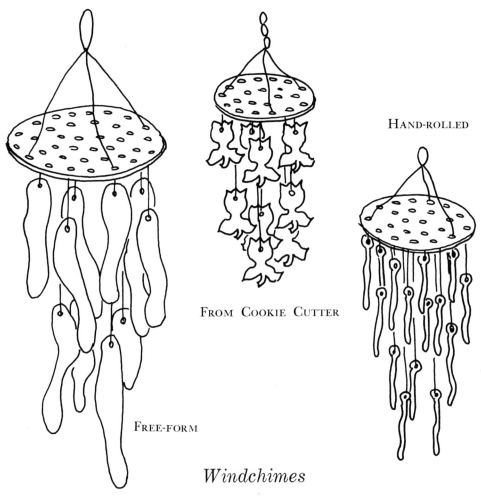

FREE-FORM

FROM COOKIE CUTTER

HAND-ROLLED

Windchimes

(See color plate.)

This simple design will give many melodious hours of enjoyment and is a good beginner's project. Hang it outside near the door or window and, rain or shine, it will chime with the breeze. Do not put it where strong winds might chip or break it. Wind chimes make a particularly nice gift for a shut-in, since weather cannot affect the glassy finish and they will last indefinitely. Use this general pattern at first; after practice, change or add to it by making it larger, smaller, narrower, wider, as you like. Change the colors as you like.

Claywork 1 hour
Dry 3–4 days
Fire 05 cone

Glaze
Fire 06 cone

MATERIALS

clay size of large apple
indelible pencil
pencil or ballpoint pen
talcum powder

ceiling tile
pointed tool
sponge
2 wooden slats 10 inches long

underglaze paints to be used with clear glaze
or colored glaze
¾ inch brush

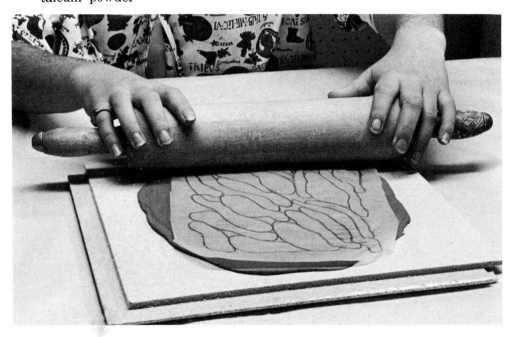

PROCEDURE

STEP 1 Place clay between slats 10 inches apart. Lightly powder roller and clay. Press roller into clay and flatten and roll up and down until clay is even and of the same thickness as the slats. Then remove slats to roll clay a little thinner.

Trace pattern A (page 37) from this text with onionskin paper, wax paper, or freehand onto the clay. Place the tracing on the clay. If you trace with indelible pencil, turn tracing over, pencil side down on the clay, and gently roll the design onto the clay. Outline tracing with pointed tool or hatpin. Remove onionskin and cut out chimes with pointed tool.

STEP 2 Cut hole in end of each chime with pencil or top end of ballpoint pen. Wipe both sides of each chime with damp sponge. Carefully place each piece on wood board or ceiling tile to dry.

When chimes are dry, smooth all rough edges with sponge. Remember that a pointed sharp edge will break easily. Check all holes to be sure they are cut completely through.

STEP 3 Roll out piece of clay between slats to a 6 x 6 inch square. Trace pattern B (page 38) onto clay. Poke top of windchime with holes as on pattern. The holes marked X are where you will attach the hanging wire.

Fire the chimes and the top to 05 cone before painting because they are so fragile. You may also paint with underglaze colors, fire again to 05 cone, apply clear glaze, and fire to 06 cone. Or coat chimes with colored glazes and fire to 06 cone.

When coating chimes with underglaze paint or glaze, hold tip end opposite the hole with fingers. Coat up to where you are holding chime, then stick point of pencil through hole and hold while finishing other end with glaze. Be sure the hole on each piece is clean of paint or glaze before you fire. When firing glazed pieces, you must use nichrome wire. Hang pieces on wire supported by kiln poles, as in the picture on page 33. Pieces must not touch during firing. The color sample was made with red, orange, and yellow glazes. A most attractive windchime may be made using blue-green, dark green, and chartreuse.

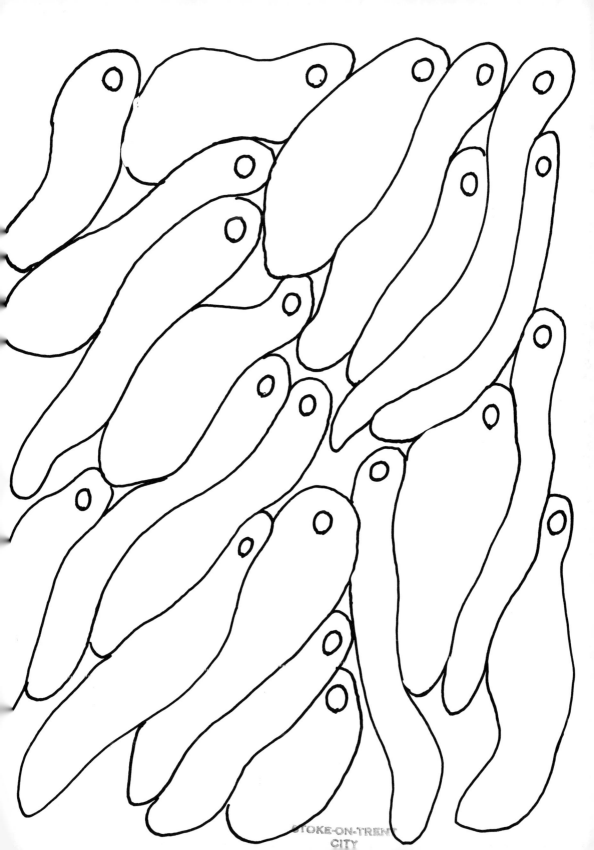

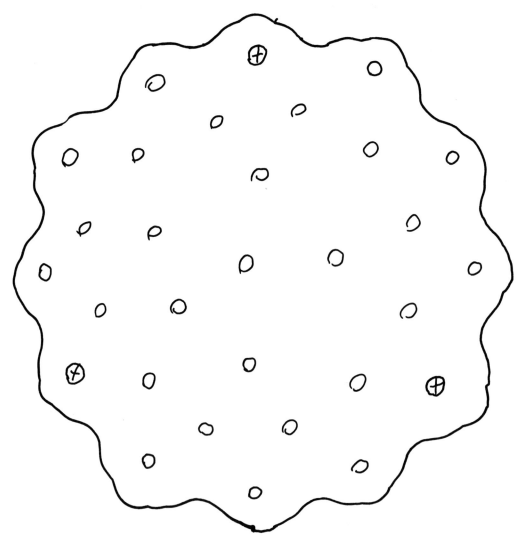

Alternate Suggestions

COOKIE CUTTER CHIMES

Cut nine or ten pieces of clay using a cookie cutter. Remember to dip the cutter in powder each time you cut so clay won't stick. Poke a hole in the top of each piece with a pencil. Clean raw edges, sponge off, dry for a week, and fire to 05 cone. Glaze and fire again to 06 cone. Yellow, orange, or blue are attractive colors to use.

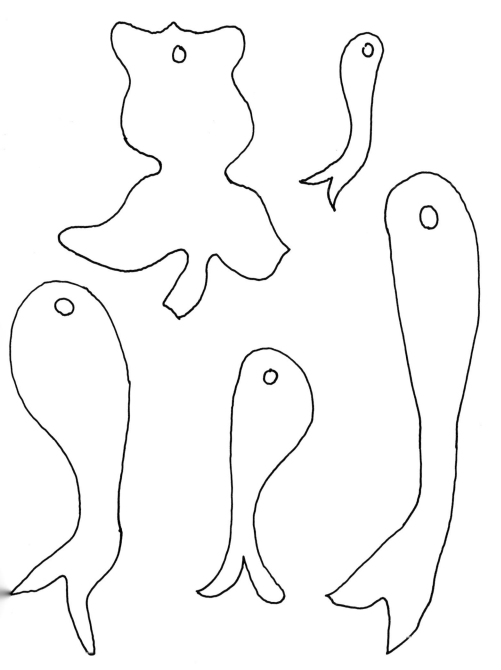

FISH CHIMES

Cut ten or twelve fish from designs shown and follow same procedure.

PROCEDURE FOR COIL CHIMES

STEP 1 Roll clay between your hands to make thin coils 3, 4, and 6 inches long.
STEP 2 Flatten each end with your fingers and poke hole through end.
STEP 3 Clean edges, sponge off, and fire to 05 cone.
STEP 4 Apply clear glaze and fire to 06 cone. Blue-green, yellow, and yellow-green, or all one color using a speckled glaze, make attractive pieces.

Large Dish with Leaf Edge

Claywork 30 minutes
Dry 5–7 days
Fire 05 cone

Glaze
Fire 06 cone

MATERIALS

clay
2 wooden slats
rolling tool
sgrafitto tool
cutting tool
sponge
5–7 leaves same size
colored glaze
brush

PROCEDURE

STEP 1 Roll clay between slats.
STEP 2 (See drawing, page 41.) Arrange five to seven leaves in a circle with the leaf veins turned down. Remove slats and roll over leaves firmly *once* to mark clay with leaf imprint. Do not repeat rolling because clay will shift and imprints will not be clear.
STEP 3 Cut clay away from outer edge of leaves. Do not make sharp points on leaf ends. Carefully lift clay from around leaves. Smooth edges with damp sponge.
STEP 4 (See drawing, page 41.) Gently place in a 6 or 8 inch casserole with flat bottom to dry for four or five days.

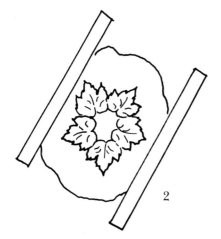

2

4

STEP 5 Carefully smooth any rough edges with damp sponge and sgraffito tool.
STEP 6 Fire to 05 cone.
STEP 7 Coat with colored glaze. Speckled green glazes are attractive.
STEP 8 Fire to 06 cone.

Oak Leaf Ashtray with Acorns

Claywork   30 minutes   Glaze
Dry   5–7 days   Fire 06 cone
Fire 05 cone

MATERIALS

clay	slip	underglaze, brown and green—with clear glaze
rolling tool	sponge	
2 wooden slats	powder	
cutting tool	screen	*or* green and brown glaze
		brushes

PROCEDURE

STEP 1 Cut pattern A, or use a natural oak leaf.

STEP 2 Roll clay between slats. Remove slats. Place pattern on clay. Roll over leaf firmly *once* so imprint will be clear. Cut around leaf with cutting tool. Smooth edges with damp sponge. You may want to make several of these at one time.

STEP 3 Make three acorns, each in two parts, by making three balls of clay the size of small grape. Pull up top stem with fingers.

STEP 4 Hollow out slightly by pushing in with fingertip.

STEP 5 While it is on the finger, roll ball on screen so as to make rough imprint on the acorn cap.

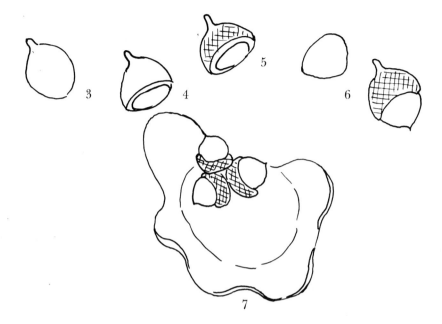

STEP 6 Make ball the size of small pea. Coat one side with slip and set inside cap.

STEP 7 Using fingers, shape leaf into dish. Score one side of acorns and place on the side of the dish to make cigarette rest.

STEP 8 Dry five to seven days, or two days' natural drying and a warm oven six to eight hours.

STEP 9 Paint leaf with green underglaze and acorns with brown underglaze.

STEP 10 Fire to 05 cone.

STEP 11 Apply three coats of clear glaze.

STEP 12 Fire to 06 cone.

Alternate

Form any shape leaf. Cut cigarette rest from pattern B. Do not attach acorns to leaf. Fire to 05 cone to make bisque. Use green glaze on leaf. Coat acorns separately with brown glaze. Put acorns in place. Fire to 06 cone. During the firing, the glaze will serve to adhere the acorns to the leaf.

*Free-form Dishes and Ashtrays
Molded over Rocks*

Claywork 30 minutes
Dry 5–7 days

Fire 05 cone
Glaze
Fire 06 cone

MATERIALS

clay
cutting tool
sponge

brush
smooth rocks (usually found near lakes or ponds)

nylon net to suit size of rock
ceiling tile
variegated glaze

PROCEDURE

STEP 1 Cover stone with net and pull tightly across surface, which will be the inside of dish. Take a ball of workable clay (clay is workable when it is dry enough not to stick to your hands and wet enough so it will not crack when molded). If clay is too wet, roll on newspaper to remove moisture. If clay is too dry, put it into plastic bag with 1 teaspoonful of water and knead.

Smooth clay over upper surface of stone. Clay may curl under stone. You will cut this away later.

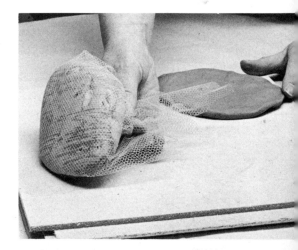

STEP 2 Pull up three or four legs. Carefully turn over dish and rest on table to make sure legs are evenly balanced. Support dish in your hands while clay is soft.

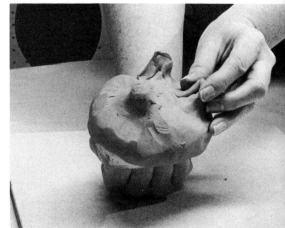

STEP 3 With cutting tool, cut evenly around stone.

STEP 4 Carefully smooth edges and clean complete dish with wet sponge. Remove dish from stone. Pull off net and clean edge and excess clay from underside. Put aside to dry.

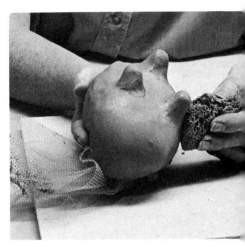

STEP 5 If to be used as ashtray, make an indentation with your finger for cigarette holder, or add a piece of curved clay.

STEP 6 Do not allow dish to dry on stone because it will crack in the shrinking process. Dry five to seven days or two days' natural drying and a warm oven six to eight hours. Clay should not be more than ½ inch thick at any point; the thinner, the better.

Fire to 05 cone.

STEP 8 Glaze with variegated glaze. Glazes containing crystals produce lovely results. Brush crystals up to edges. Do not allow crystals to pile in the bottom of dish.

STEP 9 Fire to 06 cone.

Alternate Clay Dish
(FOR BEGINNERS)

Claywork 30 minutes   Glaze
Dry 5–7 days   Fire 06 cone
Fire 05 cone

MATERIALS

clay   glaze brush   nylon net
stone   colored glaze

PROCEDURE

STEP 1   Make thirty small balls of clay.

STEP 2   Cover stone with net.

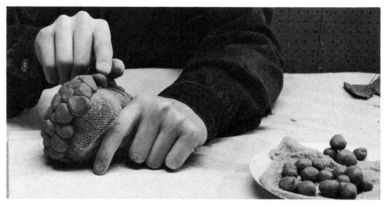

STEP 3   Cover sides and bottom of stone with the clay balls. Only cover halfway up the side of the stone with the balls so that stone can be easily removed from clay piece when finished. It is not necessary to use slip to attach balls if they are firmly pressed together. Also, that will eliminate having any large holes to fill up with glaze.

STEP 4   Add three legs to bottom, using two or three balls per leg. Set on the table on legs to make sure dish is evenly balanced.

STEP 5   Dry five to seven days, or after two days' normal drying, six to eight hours in a warm oven.

STEP 6   Fire to 05 cone.

STEP 7   Glaze with any colored glaze or a colored variegated glaze.

STEP 8   Fire to 06 cone.

Birdhouse

Claywork   1 hour
Dry   5–7 days
Fire   05 cone

Glaze
Fire   06 cone

MATERIALS

rolling tool
clay

cutting tool
paper towels or newspaper
ceiling tile

colored glaze
brush

PROCEDURE

STEP 1 *Form* Begin with a large, egg-shaped wad of paper towels or newspaper, 10 by 4 inches. Have the top layer of paper smooth and even because this layer will stick to clay when form is removed. It will burn off during firing.

STEP 2 Roll out three or four slabs of clay, 6–8 inches in diameter and ¼ inch thick.

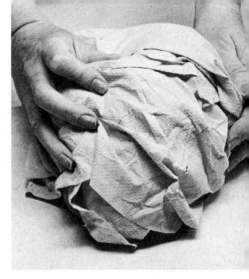

STEP 3 Place first slab over form. Score all edges and coat with slip.

STEP 4 Place second slab on form, joining to first. Continue adding slabs until form is covered. Carefully smooth complete surface with damp sponge.

STEP 5 Cut entrance to birdhouse the desired size, depending on type of bird you want to attract.

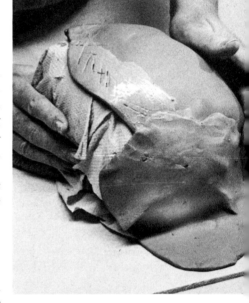

STEP 6 Cut small round hole ¾ inch at base of entrance for perch. Make a roll of clay ¾ inch in diameter, 2½ inches long. Score, coat with slip, and put perch in place. Seal joint carefully by adding small ring of clay over seam and smoothing with damp sponge.

STEP 7 Cut excess clay away from edge of house, leaving a small rim if desired.

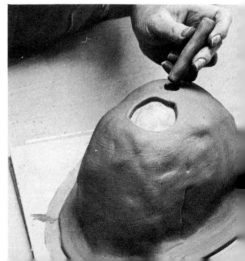

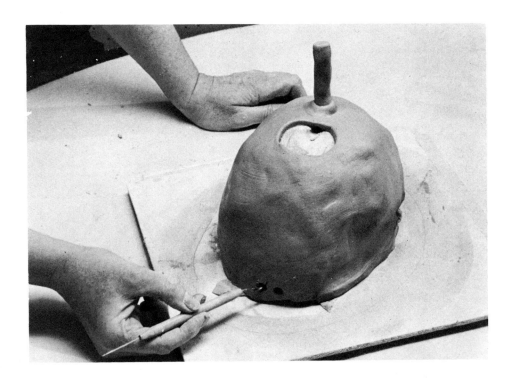

STEP 8 With pointed tool or top of ballpoint pen, put in four sets of double holes in the top, bottom, and on each side. They will be used for attaching to board when hanging.

STEP 9 Dry five to seven days, or two days normal drying and a warm oven six to eight hours.

STEP 10 Fire to 05 cone.

STEP 11 Glaze green speckled or wood tone. Glaze inside of house, too, to aid cleaning.

STEP 12 Fire to 06 cone.

STEP 13 String wires through holes and attach to tree or board. Birdhouse can then be easily removed for cleaning.

Clown Bank

(See color plate.)

Banks can be made in all shapes and sizes. Mold them around paper balls. Make them in shapes of animals, fun characters or make a clown bank.

Claywork 1 hour
Dry 5–7 days
Fire 05 cone
Underglaze

Fire 06 cone
Glaze with clear matt or gloss
Fire 06 cone

MATERIALS

clay
paper towels or newspaper
rolling tool

underglaze paints—
 red, black, yellow, blue
glaze brushes
glaze

wood slats
ceiling tile
sponge

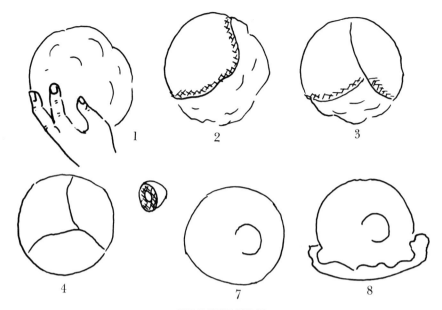

PROCEDURE

STEP 1 Take two 12-foot-long pieces of paper toweling. Make them into a tight ball the size of a large apple.

STEP 2 Take four small balls of clay each the size of a large plum and roll between slats into flat strips ¼ inch thick. They should be 4–5 inches wide. Wrap first strip of clay around paper ball. Score all edges of strip and coat with slip.

STEP 3 Wrap second strip of clay around paper ball, overlapping first strip. Score top edge of second strip, and add third and fourth strip in same manner.

STEP 4 Wrap third piece of clay around paper ball, overlapping scored edges, which have been coated with slip.

STEP 5 Wrap fourth strip of clay around paper ball. This should completely cover the ball.

STEP 6 Smooth seams with damp fingers and wet sponge.

STEP 7 Make another ball the size of small plum. Flatten on one side. Hollow out center. Score edge, coat with slip, and attach for nose of clown.

STEP 8 Roll out piece of clay the size of small apple between slats to ¼ inch thickness. Cut collar from pattern A. Turn edge up with fingers. Score bottom of head and top of collar. Coat with slip and place head on collar, pressing down firmly.

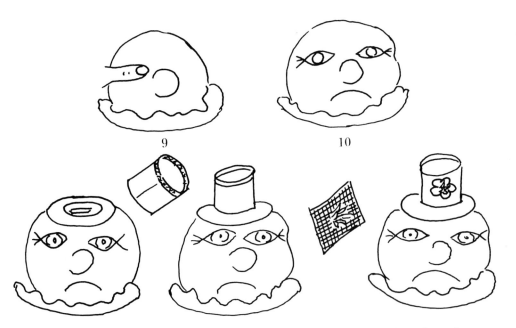

STEP 9 Push in gently above nose on each side to make place for eyes.

STEP 10 With pointed cutting tool, scratch eyes in hollows and cut in mouth as shown. Scratch long marks across back of head to show hair.

STEP 11 Cut small hole the size of a quarter in bottom. Do not try to remove paper. It will burn away in the firing.

STEP 12 Roll out clay and cut hat and brim from patterns B and C. Smooth edges with damp sponge. Score top of clown head. Coat with slip and place hat rim (pattern C) on head.

STEP 13 Cut hole the size of quarter in the top, where hat is going to be.

STEP 14 Score edges of hat top (pattern B). Coat with slip and join to make a cylinder. Smooth seam with fingers and damp sponge.

STEP 15 Score edges on bottom side of hat cylinder. Coat with slip and attach to hat brim. Hat should be left hollow so coins can fall through.

STEP 16 Cut flowers from pattern D. Thin edges of flower petals by gently pressing with fingers.

STEP 17 Push clay through screen to form flower center. Score hat and coat with slip where flower is to be attached.

STEP 18 Attach flower to hat. Coat center of flower with slip and attach flower center.

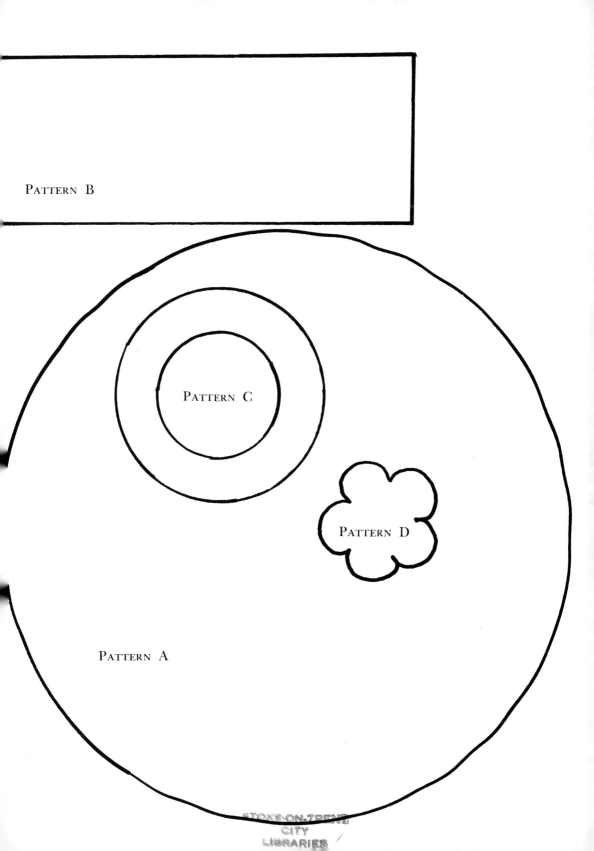

STEP 19 Clean complete piece with very damp sponge. Do not remove paper. Dry for seven to ten days. If clay is very thick, place in low-heat oven for three to four hours at end of drying period. You may paint before firing—this may be best for a beginner because of easier handling.

STEP 20 Paint with underglaze paints (takes about an hour).

Flower petals, nose, mouth—red
Inside diamonds around eyes—red
Circle around eyes—blue
Center of eyes—blue
Eyebrows—black
Hat—black
Whiskers—black
Outline around eyes—black
Flower center—yellow
Collar—yellow
Hair—green

STEP 21 Fire to 05 cone.

STEP 22 Coat with three thin coats clear glaze. Fire to 06 cone.

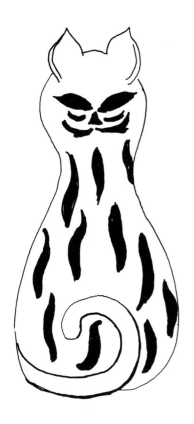

Cat Patio Light

Claywork 1 hour
Dry 6–24 hours
Cut light holes 30 minutes
Dry 5–7 days

Fire 05 cone
Glaze
Fire 06 cone

MATERIALS

12 oz., 10 inch soda bottle clay

rolling tool
sgrafitto tool
newspaper or paper towels

pointed cutting tool
sponge

STEP 1 Make a tight paper ball the size of large apple. If ball is not very tight, the weight of the clay head will push down over the bottle and mar the shape. Push top of soda bottle halfway up into paper ball.

STEP 2 Wrap the rest of bottle with paper to form neck and fat body. This is the form on which to mold the cat. The bottle will be removed from the form later. Do not cover the bottom of bottle with paper. Place form on small board or piece of ceiling tile.

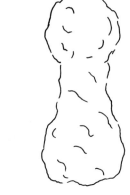

STEP 3 Using rolling tool and slats, roll out five or six small pieces of clay the diameter of a spread hand (approximately 6 inches across).

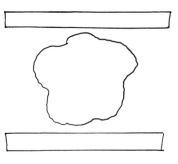

STEP 4 Roll circle of clay approximately 8 inches in diameter. Place over head of cat. Clay will fold in some places. Cut away excess. Smooth over with fingers. If clay is as much as ¾ inch thick in places, it will dry and fire satisfactorily.

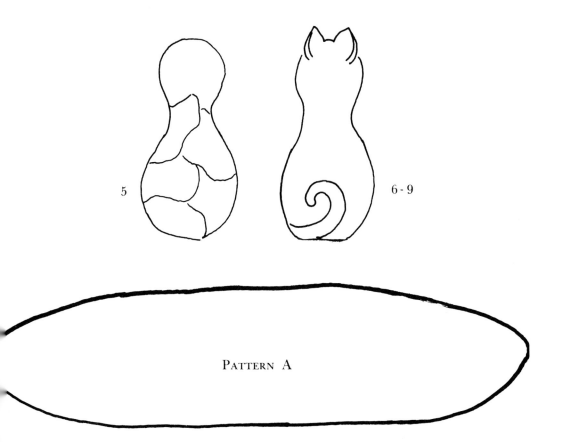

STEP 5 Score all edges of head piece. Coat with slip. Take clay pieces and cover rest of form. Smooth and seal each seam as you work. Use slip freely. If clay separates, coat with slip and patch.

STEP 6 Roll ¼ inch thick piece of clay between slats. Using pattern A, cut out ears. Score and coat top of head with slip. Place ears on top of head. Keep ears forward on head. Shape and smooth the ridges of clay. This will also give top of head strength and shape.

STEP 7 With a dripping-wet sponge, carefully smooth seams on body, removing all pitmarks.

STEP 8 Roll long, thin coil approximately 12 inches long, for tail.

STEP 9 Flatten one end. Score, coat with slip at back on bottom of cat and around to front of cat, where tail is to be attached. Attach tail firmly to body, tucking flat end of tail underneath cat at center back.

11

STEP 10 Dry several hours or overnight, depending on climate. Dry until leatherlike and easily cut. It is considered easily cut when clay does not crack or stick to cutting tool.

STEP 11 Cut out eyes and whiskers and oval holes with pointed tool as shown. Smooth over with damp sponge and put aside to dry five to seven days, or two days' normal drying and six to eight hours in a warm oven.

STEP 12 When completely dry, clean around holes with sgrafitto tool and smooth with damp sponge. Remove bottle and loose paper. Paper that sticks will burn away in the firing.

STEP 13 Fire to 05 cone. Glaze with gloss black. Do not put glaze inside. Do not put glaze on bottom. Clean excess glaze from bottom with sandpaper, so that when it is fired, it will not be necessary to set it on stilts. You may set it flat on kiln bottom. Glaze inside with clear glaze or do not glaze inside at all. A white interior throws out more light.

STEP 14 Fire to 06 cone.

Make a simple candle-holder to hold small candle inside patio light. See page 64.

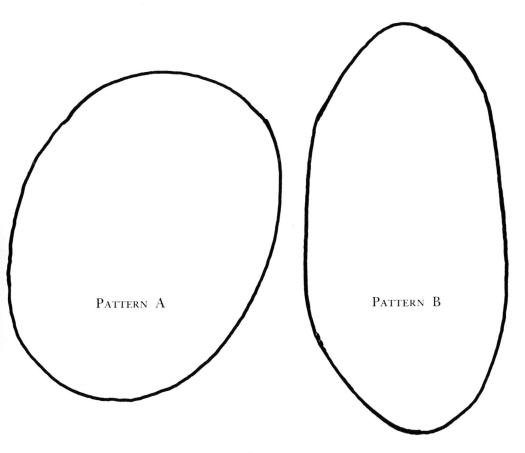

Owl Patio Light

(See color plate.)

Claywork 1 hour
Dry 6–12 hours
Cut out light holes
   30 minutes
Dry 5–7 days
Fire 05 cone
Glaze
Fire 06 cone

MATERIALS

clay
rolling tool

12 oz., 8 inch soda bottle
sponge
newspaper or paper towels

pointed cutting tool
sgrafitto tool

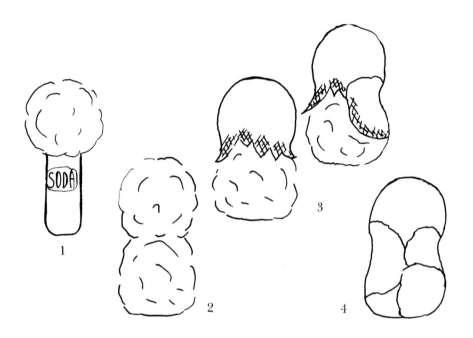

PROCEDURE

STEP 1 For the head, make a tight paper ball the size of small melon. Press down over neck of bottle.

STEP 2 Wrap rest of bottle tightly with paper. This is the form on which to mold the owl. Do not cover bottom of bottle.

STEP 3 Roll six pieces of clay the size of a spread hand. Cut one piece about 10 inches wide. Place 10 inch piece over head and shape to form head. Cut away excess clay where it folds and score around bottom edge.

STEP 4 Attach additional pieces of clay over rest of form, scoring all edges that overlap. Smooth all seams with wet fingers.

STEP 5 Cut two clay pieces of pattern A approximately $\frac{1}{2}$ inch thick. Attach on front of head to form jowls. Score each side of head and coat with slip where jowls are to be attached. Press very firmly on jowls to attach and then press with thumb to form eye sockets. Pinch out bridge of nose. Add small ball of clay for beak.

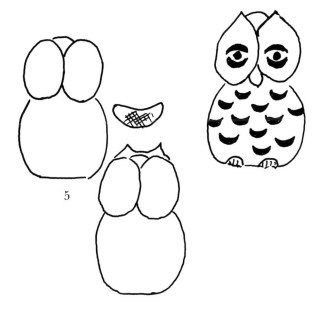

5

STEP 6   Flatten top of head. Cut pattern B for ears ½ inch thick. Score top of head and coat with slip. Attach ears. Carefully pinch up ear ends and form ridge to meet nose.

STEP 7   Make two balls of clay the size of grape. Score, coat with slip, and attach under front of owl for claws. With end of brush handle, make several marks to show claws.

STEP 8   With dripping wet sponge, smooth over complete owl to fill in pit holes and eliminate seams. Smooth out eye sockets and jowls. Take time to do a complete smoothing job. If there are deep pit holes, the glaze will not cover.

STEP 9   Put aside for six to twelve hours until leatherlike and easily cut without sticking to cutting tool.

STEP 10   Cut out eyes and holes in body as shown in drawing. Smooth with damp sponge. Remove bottle and any loose paper. Put aside to dry five to seven days. Carefully clean with sgrafitto tool if necessary. Remnants of paper will burn off in the firing.

STEP 11   Fire to 05 cone. Glaze with speckled brown, orange, or gold. Fire to 06 cone. Do not apply colored glaze inside. Use clear glaze or leave white bisque.

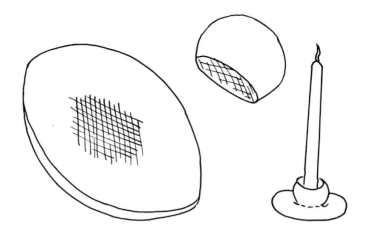

Candle-holder for Patio Light

Claywork 30 minutes
Dry 5–7 days
Fire 05 cone

Glaze
Fire 06 cone

MATERIALS

clay   clear glaze, if desired   candle   brush

PROCEDURE

STEP 1 Take ball of clay the size of plum. Flatten in hand. Smooth edges. Make sure it is small enough to fit inside patio light. It should be approximately 2½ inches in diameter. Score middle of top surface. Place on ceiling tile.

STEP 2 Take another ball of clay the size of small plum. Flatten the underside. Score and coat with slip. Press down on the flat piece in the center to attach firmly.

STEP 3 Press candle base firmly into ball far enough so holder will support candle. Rotate candle slightly to make hole large enough to allow for shrinkage of the clay during firing—about ¼ inch larger than candle.

STEP 4 Dry five to seven days, or two days' natural drying and a warm oven six to eight hours.

STEP 5 Fire to 05 cone. You may stop here without glazing since this piece will not be seen, or you may apply clear glaze and fire to 06 cone. You can also make matching candle-holders to go with the leaf dish.

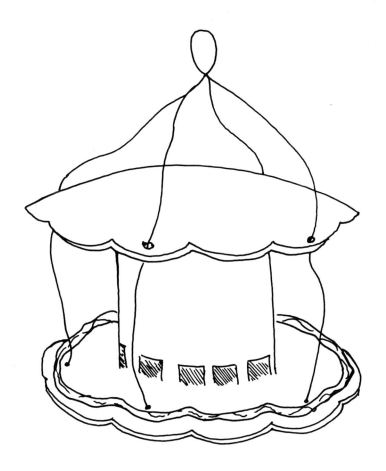

Bird Feeder

(See color plate.)

Claywork 2 hours
Dry 5–7 days
Fire 05 cone

Glaze
Fire 06 cone

MATERIALS

3 pounds clay
rolling tool
2 wooden slats
Scotch tape
oatmeal box

pencil
wax or tracing
 paper
ceiling tile
slip

cutting tool
sponge
yellow glaze
¾ inch brush

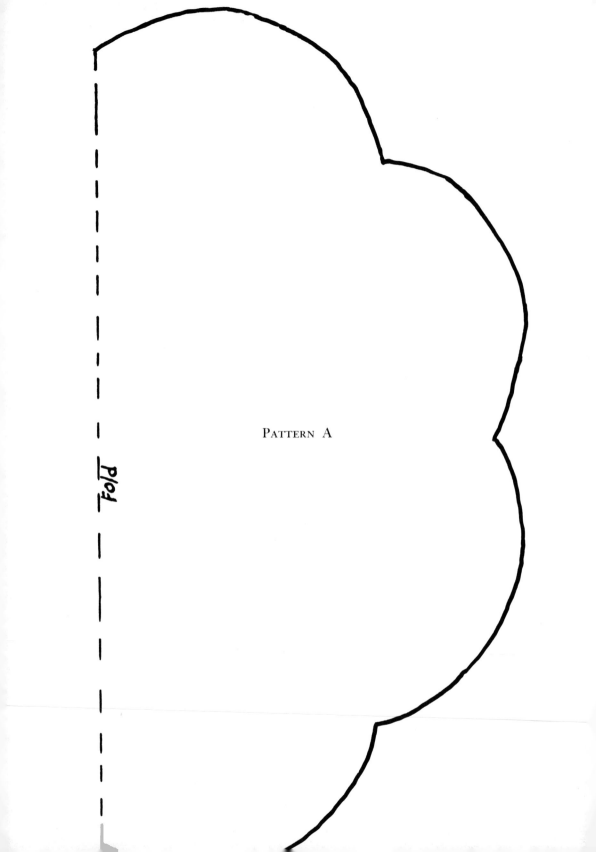

PROCEDURE

STEP 1 Trace pattern A from text onto wax or tracing paper. Only half of the pattern is shown. Fold at dotted line.

STEP 2 Duplicate pattern onto cardboard for easier handling, particularly if it is to be used by several people.

STEP 3 Take piece of clay the size of grapefruit. Roll between slats 10 inches apart until clay is smooth and same thickness as slats. Place base pattern A on clay and cut with cutting tool. Hold pattern firmly on clay. Lift and remove clay from around base. Clean base with damp sponge to remove rough edges. Place base on large wooden board or on ceiling tile for easier handling. When feeder is finished, keep on ceiling tile for a week while drying. Save pattern. It will be used again for the lid.

STEP 4 Take a round oatmeal box and cut out bottom and top. Carefully cut down one side but do not spread box open.

STEP 5 Tape box together from inside so it will hold shape. Set box in middle of base. Score clay around base of box with cutting tool. Coat scoring with slip.

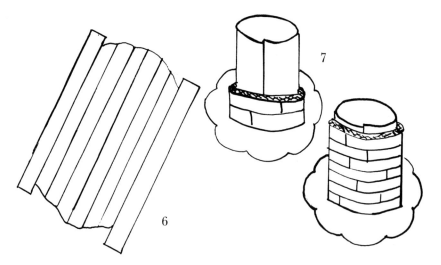

STEP 6 Between wooden slats roll out piece of clay the size of grapefruit to a length of 14 inches. This is the length necessary to make clay strips to encircle the average oatmeal box. If you use a different size box, measure distance around box with string to get necessary length of strips. For an oatmeal box, cut six strips of clay approximately ¾ inch wide and 14 inches long. Strips must be straight. Place ruler on the clay and cut with pointed tool.

STEP 7 Place first clay strip around base of box. Score both edges of strips with small crosses and coat with slip. Score strip ends and join together with slip. Do not join ends of succeeding strips in the same spot as previous ones. Join at different points, as shown. Apply second strip in same manner.

STEP 8 Score edges and place strips one above the other, making sure all scoring is covered with slip before placing next strip.

STEP 9 Smooth all joinings with wet fingers, leaving box in place. Wipe outside surface with damp sponge to erase all marks or pin holes.

STEP 10 *Perch* Take piece of clay the size of apple. By rolling clay between your hands, make two strips of clay 12 inches long and the width of a finger. Strips do not have to be uniform or smooth because the roughness will give better footing for the birds.

STEP 11 Score outer top edge of base with pointed tool. Coat with slip. Press clay strips around base to make perch. Score ends of strips. Coat with slip and join. Wipe over with damp sponge.

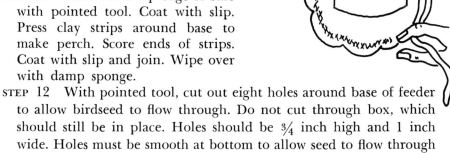

STEP 12 With pointed tool, cut out eight holes around base of feeder to allow birdseed to flow through. Do not cut through box, which should still be in place. Holes should be $3/4$ inch high and 1 inch wide. Holes must be smooth at bottom to allow seed to flow through easily. Wipe over holes with damp sponge to remove all rough edges.

STEP 13 Cut four small holes in bottom slab, to attach hanging wire. Holes must be evenly placed so bird feeder will hang evenly. Holes should be as close to perch as possible and should be on inner side of perch.

STEP 14 *Lid* Use pattern A again. Take piece of clay the size of grapefruit and roll between slats until clay is same thickness as slats. Place pattern A on clay and cut lid with pointed tool. Wipe edges smooth with damp sponge.

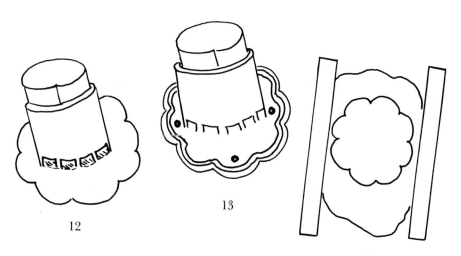

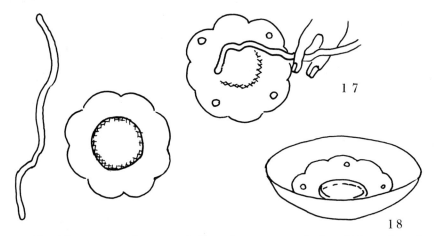

STEP 15 Place top of oatmeal box in center of clay lid and twist gently to mark circle in center of clay. Do not cut clay with top. Using pointed tool, score inside circle mark on clay.

STEP 16 Roll piece of clay between hands to make rope 11½ inches long and the width of a finger.

Apply slip on scoring.

STEP 17 Place rope on lid over scoring. This will keep lid in place when piece is finished. Cut holes in four places for hanging wire, as shown.

STEP 18 To make lid slightly concave, place it inside shallow bowl or plate to dry, keeping the rope side up. This must dry five to seven days before firing.

STEP 19 Remove box from base by removing Scotch tape and squeezing box together gently. Coat inside of base thoroughly with slip to smooth all seams. With damp sponge, carefully clean up all rough edges on bird feeder.

STEP 20 Dry five to seven days or two days' natural drying and a warm oven six to eight hours.

STEP 21 Fire to 05 cone.

STEP 22 Apply three coats of green glaze (yellow or orange are attractive, too). Be sure to glaze all surfaces including bottom since the feeder will be used outside. Glaze and fire the lid separately.

STEP 23 Fire to 06 cone.

STEP 24 To assemble, cut two pieces of rustproof wire, 40 inches long. Run each piece under bottom of feeder and up through two holes. Put wire ends through two holes on lid and bring the four pieces to a center point. Twist and tie in a loop. The lid will easily slide up and down for quick filling with birdseed.

Angel Figurines for Christmas

Claywork 1 hour
Dry 5–7 days
Underglaze
Fire 06 cone

Glaze
Fire 06 cone
Trim with liquid gold
Fire 018 cone

MATERIALS

clay the size of apple
2 wooden slats
lightweight cardboard

brush
cutting tool
screen
rolling tool

ceiling tile
underglazes
glaze

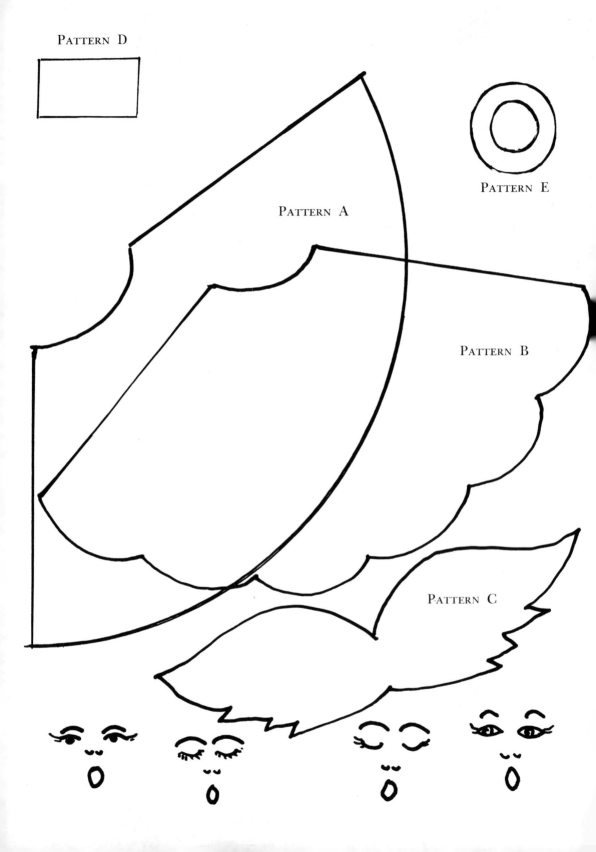

PROCEDURE

STEP 1   Roll out clay between slats ¼ inch thick. Cut pattern A from cardboard. Form into cone and Scotch tape together from inside. This is cone on which you build angel.

STEP 2   Cut patterns B, C, and D from paper.
Cut body pattern B ¼ inch thick.

 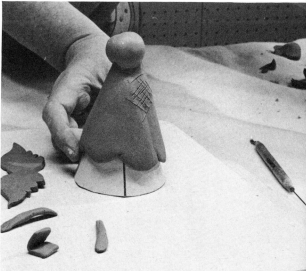

STEP 3   Cut clay wing of angel ⅛ inch thick, pattern C. Form body pattern into cone. Score edges. Coat with slip and smooth seam with fingers. Place clay body cone over the cardboard cone.

STEP 4   Make ball of clay the size of small plum. Hollow out with thumb and put on clay body. This is the head.

STEP 5   With cutting tool, put three holes through head to allow air to escape. Holes will be covered by hair later.

STEP 6   Cut pattern D for book. Shape book into open position with fingers.

STEP 7   Roll out two arms about 3 inches long and ½ inch around.
Flatten one end for hands. Mark in fingers with pointed tool.

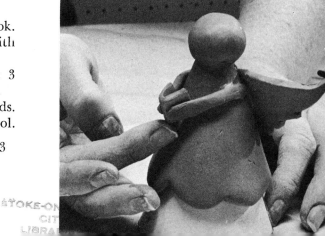

STEP 8   Score back of angel at shoulders. Coat with slip and flatten ends of arms and attach to body. Shape arms and hands into position and place book between hands. Or, if desired, angel can hold small candle. Take ball of clay the size of small grape. Press in with finger so it will hold small candle, and attach to hands in same manner as book.

STEP 9   Smooth edges of wings with damp sponge and fingers and bend wings back slightly. Score back of angel, coat with slip, and attach wings.

Score and coat with slip top of head, where hair will be placed.

STEP 10   Take piece of clay the size of a pea. Push through screen. Lift this clump of hair with pointed cutting tool and place on head. Make another clump and repeat, until head is completely covered.

STEP 11   Flatten $1/8$ inch thick piece of clay. Cut halo from pattern E. Smooth with fingers and place on head. Allow completed angel to dry on cone.

STEP 12   Dry five to seven days, or after two days' normal drying place in a warm oven for six to eight hours.

STEP 13   Fire 05 cone.

STEP 14   Paint with flesh-color underglaze first on face. Draw faces with black underglaze. Underglaze paints—yellow hair, gold halo, blue book, blue trim on dress. If you paint faces over unfired flesh underglaze, you can easily wash off face with damp sponge if you make a mistake.

STEP 15   Fire 06 cone.

STEP 16   Glaze with clear matt glaze.

STEP 17   Fire 06 cone. If desired, trim halo and wings with liquid gold, which must be applied over the fired glaze.

STEP 18   Fire 018 cone. Note special cone for gold and metallic finishes.

Three Wise Men Christmas Figurines

Claywork each figurine 1½ hours
Dry 5–7 days

Fire 05 cone
Glaze
Fire 06 cone

MATERIALS

clay
rolling tool
2 wooden slats

screen
cutting tool

slip
brushes
glazes and underglazes

PROCEDURE FOR FIRST WISE MAN

STEP 1 To make a cardboard cone to support the figurine, trace solid line design of pattern A (page 76) onto tracing paper and place on piece of lightweight cardboard. Cut out and form into cone. Seal joining with Scotch tape on inside of cone. (It is easy to remove tape and cone when figurine is dry.)

STEP 2 Roll out clay the size of an apple between wooden slats. Trace broken-line design of pattern A onto wax or tracing paper. Place pattern on clay and cut with pointed cutting tool.

STEP 3 Score and coat seams with slip and smooth carefully with fingers. Clay cone should slip easily over the cardboard support cone. The cardboard should show approximately ¼ inch below the clay. This is to prevent marring the bottom edge of the clay before it is dry.

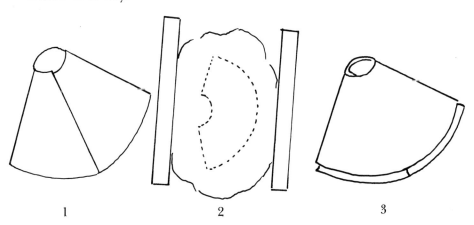

1 2 3

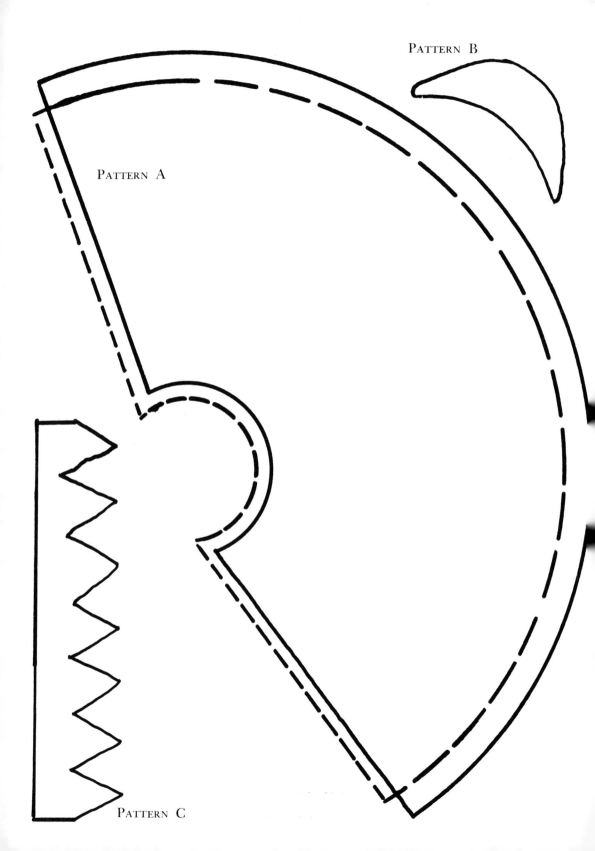

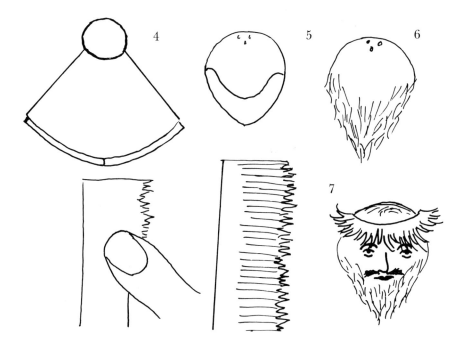

STEP 4 *Head* Form clay ball the size of grape. Using thumb, hollow out inside. With a pointed tool, make two or three holes through top of head from inside out. This will allow air to escape during firing. The holes will be covered with the hair. Score top of clay cone, coat with slip, and attach head to cone body. Finished head should be the size of a walnut. Smooth head seams at body.

STEP 5 *Beard* Cut pattern B from clay 1/8 inch thick. Score underside, coat with slip, and attach to head. Smooth seam of beard joining head.

STEP 6 Using small brush handle, mark beard with deep rough strokes. Do not cut all the way through.

STEP 7 *Hair* Cut clay piece 1/8 inch thick, 3/4 inch wide, and 4 inches long. Place flat on newspaper and thin one edge by pulling clay out with fingers.

STEP 8 Using cutting tool, cut through piece to make strands. Lift and curl hair with fingers. Place around head, keeping strands down. Use leftover strands to lie loosely across top of head, which will cover the holes. Do not plug holes.

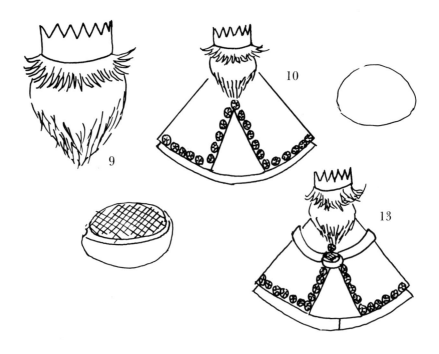

STEP 9 *Crown* Cut crown from pattern C, 1/8 inch thick clay. Carefully pinch points to thin edge. Score, coat with slip, and place over hair around head.

STEP 10 *Robe* To mark in robe, make a V up front of robe with brush handle. With fancy button or earring, press design on each side of V and around bottom of robe.

STEP 11 *Arms* Roll clay between hands to make two arms approximately 3 inches long. Flatten ends of arms, score, coat with slip, and attach arms to back of cone. Bring arms around so they are about 1 inch apart in front.

STEP 12 *Gift Box* Take piece of clay the size of small grape. Roll into ball. With brush handle, make circular indentation to resemble gift box lid as shown. With a piece of screen or fancy button, imprint design on top of box.

STEP 13 Mark hand ends of arms to resemble fingers. Coat with slip and place box between hands.

STEP 14 Dry five to seven days, or two days' natural drying and a warm oven six to eight hours.

STEP 15 Fire 05 cone.

STEP 16   Paint with underglaze paints.
Face and hands—flesh   Robe—purple
Crown—gold   Trim on design—dark blue, black
Gift box—gold   Hair and beard—brown
STEP 17   Fire to 05 cone.
STEP 18   Glaze clear matt. Do not glaze face, hands, hair, beard.
STEP 19   Fire to 06 cone.

PROCEDURE FOR SECOND WISE MAN

This wise man will be taller than the first.

STEPS 1–6 as in the first wise man. Use pattern D (page 80) for cardboard cone and body.

STEP 7 *Hair*   Push clay through screen to make hair strings. Score head in circle, coat with slip, and attach ring of hair around head.

STEP 8 *Turban*   Roll coil of clay between hands same thickness as shown. Score head, coat with slip, and wind coil around head in successively smaller circles as shown. Be sure to leave small hole at top to let air escape.

STEP 9 *Arms*   Roll clay betwen hands to make arms, each 3½ inches long. Score ends of arms and cone, coat with slip at joining, and attach arms to back of cone. Bring arms around so they are approximately ½ inch apart in front.

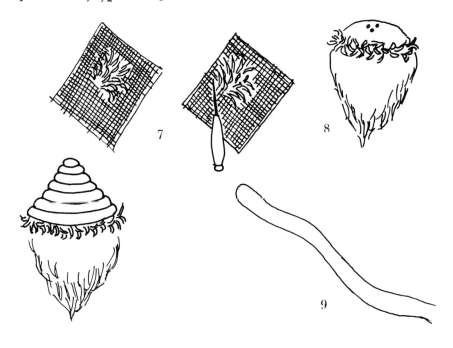

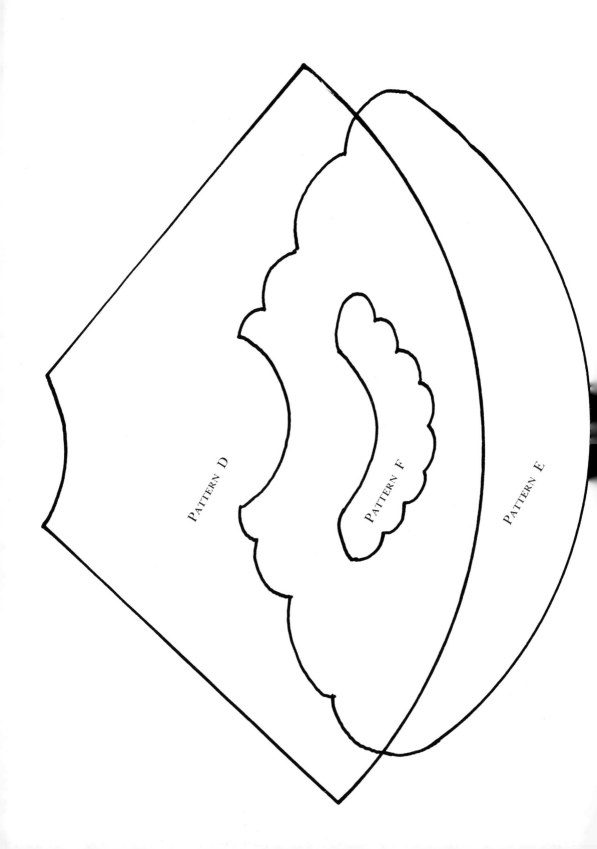

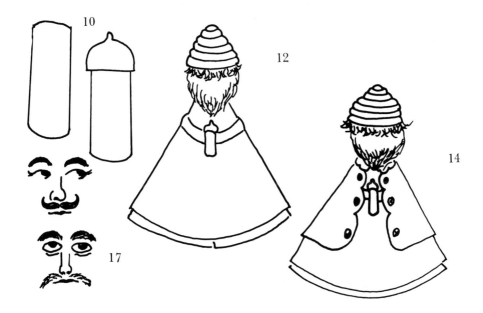

STEP 10 *Bottle* Roll piece of clay the size shown. Form a cap. Score bottle, coat with slip, and place cap on bottle.

STEP 11 Score bottle and hands where they are to join. Coat with slip and attach hands to bottle.

STEP 12 *Cloak* Cut pattern E in ⅛ inch thick clay. Score neck of clay cone and cloak. Coat with slip and drape cloak around cone body, attaching cloak to neck.

STEP 13 Cut collar ⅛ inch thick from pattern F. Score cloak and collar. Coat both with slip and place collar over cloak.

STEP 14 *Buttons* Form six little balls smaller than a pea. Flatten to resemble buttons. Score, put dot of slip on each scallop, and attach buttons. Mark center of each button with an X to decorate.

STEP 15 Dry five to seven days, or after two days natural drying, place in a warm oven six to eight hours.

STEP 16 Fire to 05 cone.

STEP 17 Paint with underglaze.

Dark blue on hat, collar, buttons and bottle
Gray on skirt
Gold on cape
Brown on hair
Flesh on face and hands. Paint face detail in black as shown.

STEP 18 Fire to 06 cone.

STEP 19 Glaze with white matt.

STEP 20 Fire 06 cone.

PROCEDURE FOR THIRD WISE MAN

Follow Steps 1–8.

STEP 9 Form bottle as shown. Coat hands with slip and attach hands to each side of bottle.

STEP 10 *Cape* Roll clay ⅛ inch thick and cut cape from pattern H. Drape cape as shown folding ends back to form collar. Score and coat underside of cape with slip so it will stick to body.

STEP 11 Cut mantle from pattern G. This should be ⅛ inch thick. Score and coat head with slip and drape around head as shown.

STEP 12 Take piece of clay the size of grape. Push in end with finger to make hat. With cutting tool, make three holes through top of hat to allow air to escape during firing. Texture hat as shown. Place hat on head to meet edge of mantle or overlap slightly.

STEP 13 Set aside to dry overnight before removing cone. Dry five to seven days.

STEP 14 Fire to 05 cone before painting because they are so fragile.

STEP 15 Paint third wise man with underglaze.

Skirt and mantle—gold   Beard—brown
Cape and hat—green   Bottle—green
Hands and face—flesh, face detail
  with black

STEP 16 Fire to 06 cone.

STEP 17 Glaze with clear matt glaze.

STEP 18 Fire to 06 cone.

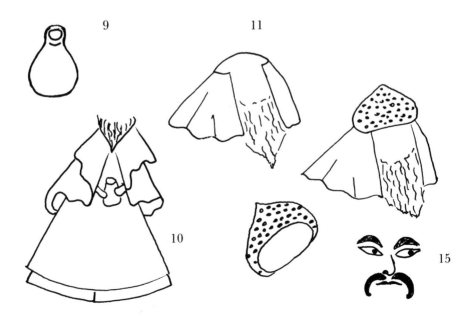

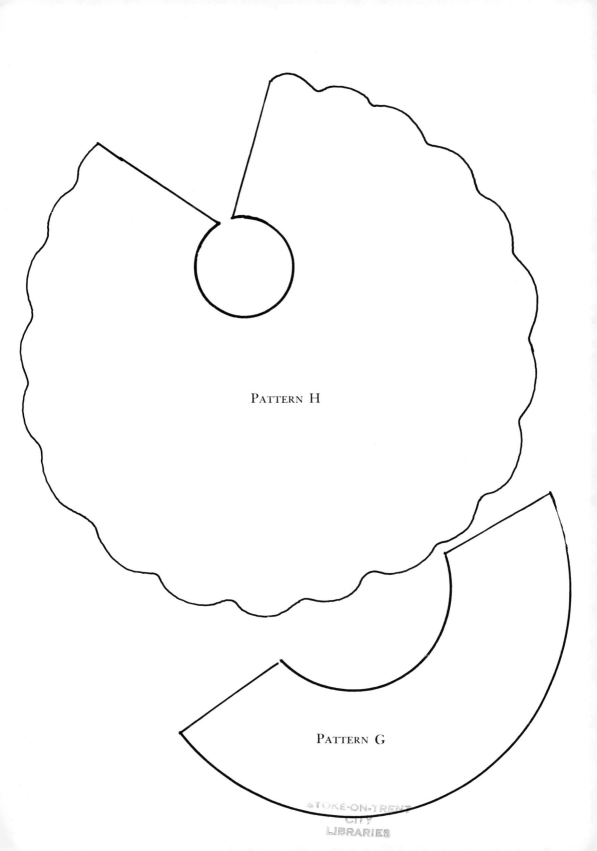

Pansy Ring

(See color plate.)

Claywork 1 hour
Dry 5–7 days
Fire 05 cone

Glaze
Fire 06 cone

MATERIALS

rolling tool
clay
2 wooden slats
sponge
cutting tool

newspaper or paper towel
6 small leaves, or use pattern
6–8 inch casserole

glass tumbler for center
green speckled glaze
brush

PROCEDURE

STEP 1 Line casserole with newspaper. Cover tumbler with newspaper and place in center of casserole. There should be a space of about 2 inches between tumbler and outer rim of casserole.

STEP 2 To make paper pattern for clay insert: Trace around bottom of bowl on paper. Trace also around tumbler so pattern resembles doughnut. Pattern should fit into casserole. Roll clay the size of apple between wooden slats. Cut clay from pattern. Put clay insert into casserole. Roll out second piece of clay between slats and cut seven leaves for outer edge, or more if needed.

STEP 3 Place leaves on clay, vein side down. Roll over leaves firmly *once* to make imprint. Do not repeat rolling or imprint will distort. Cut around leaf with cutting tool. Lift clay from around leaves. Smooth edges with fingers and damp sponge.

STEP 4 Score outer edge of clay insert. Coat with slip. Place leaves around outer edge of casserole. Leaves should overlap scoring. Keep vein side down. Do not fold leaf tip down at this time. Leaves will stand up at side of casserole. They should overlap each other approximately 1/4 inch so there will not be any holes in bottom, where water will be. Where leaves overlap, score and coat with slip. Otherwise, as clay shrinks in drying process, leaves will separate.

STEP 5 Cut five more leaves. Score and coat with slip on inner edge of insert. Place leaves around tumbler, vein side against tumbler. Be sure to score and cover with slip where leaves overlap.

STEP 6 With wet sponge, carefully smooth inside of ring. If there are any holes in the water area, score, coat with slip, patch, and smooth.

STEP 7 Twist a long piece of newspaper or toweling to put inside ring to support leaf tips. Carefully bend tips over. They may meet but should not overlap too much. Remove inside tumbler but do not remove clay ring for twenty-four hours.

STEP 8 Dry five to seven days, or two days' natural drying and a warm oven six to eight hours.

STEP 9 Fire to 05 cone.

STEP 10 Apply speckled or fancy green glaze.

STEP 11 Fire to 06 cone.

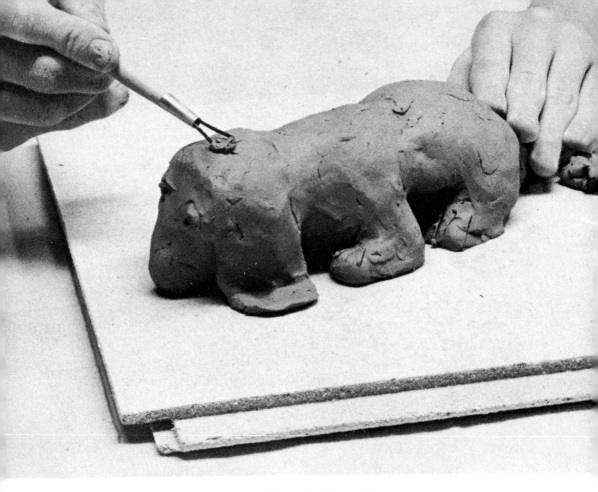

Hand Modeling

Hand modeling is the most creative type of expression in claywork. As a start, it is good to have a picture or a small object to use as a model.

Before beginning the project, cut through the clay with wire to remove air pockets. The hand-molded piece will be thicker when fired and must be free of air pockets. Throw the clay on the floor onto newspaper and cut through again. If the clay sticks to hands, it has too much water in it. Roll it on newspaper until easily worked without sticking. If clay dries too quickly, wet your fingers occasionally as you work with it. (See General Procedure chapter, pages 1–7.)

Clay dries differently at different temperatures. The thicker the piece, the longer the drying and firing times. After the piece is modeled, the inside must be hollowed out so that clay is no more than 1 inch thick at any point. This will allow the piece to dry and fire without exploding in the kiln.

It is better to pull out arms, legs, or ears from the piece rather than adding bits of clay. Too often air will become trapped under a piece that is added, and it will separate in the firing. If it is necessary to add clay, be sure to score, coat with slip, and seal seams. A lace tool is useful for smoothing or rounding the piece as desired.

Animals, heads of various kinds, or figurines make interesting subjects for easy modeling. Be sure the pieces are hollowed out to a 1 inch thickness and dried for two weeks (or placed in a warm oven for six to eight hours after one week of natural drying). Then the pieces may be fired, painted, fired, and glazed as desired.

Bookends

Claywork 30 minutes
Dry 7–10 days
Glaze

Fire 05 cone
Dry
Fire 06 cone

MATERIALS

rolling tool
clay
2 wooden slats

sponge
cutting tool

comb
ceiling tile
wood glaze

PROCEDURE

STEP 1   Cut cardboard pattern for bookends from pattern A.
STEP 2   Roll clay ½ inch thick, using double slats on each side. Cut clay from pattern A. Make two. Turn clay over and smooth underside with damp sponge.
STEP 3   On top side of each bookend, pinch edge with fingers carefully.
STEP 4   Scratch top surface lightly with comb.
STEP 5   Carefully place on ceiling tile to dry in finished shape with another tile perpendicular between two bookends.
STEP 6   Dry seven to ten days.
STEP 7   Fire to 05 cone.
STEP 8   Glaze with wood glaze. Place finished horse heads, painted and ready for glaze fire (see page 90), on bookends. Pieces will attach in the firing process.
STEP 9   Fire to 06 cone.

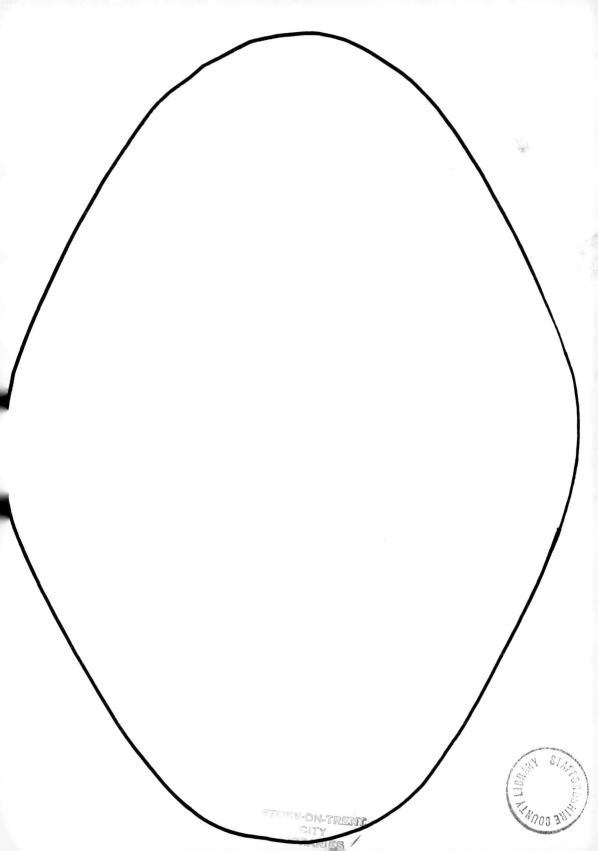

Horse Head

Claywork   30 minutes
Dry   14 days

Fire 05 cone
Glaze
Fire 06 cone

MATERIALS

clay   cutting tool   underglaze paints

PROCEDURE

STEP 1   Use piece of clay the size of small apple. Press flat in oval shape. Pull up into cone shape. Push in front and pull top over to form neck and head.
STEP 2   Pull out ears.
STEP 3   With cutting tool, mark in mane.
STEP 4   Make small hole for eyes and insert small ball of clay for eyes if desired.
STEP 5   Cut mouth.

STEP 6   When finished, turn upside down and hollow out center and neck (not head), to 1 inch thickness. This will help the drying process; pieces over an inch thick may explode when fired.
STEP 7   Dry fourteen days.
STEP 8   Paint with underglaze colors.
STEP 9   Fire to 05 cone.
STEP 10   Glaze.
STEP 11   Fire to 06 cone.

Dollhouse Food

(See color plate.)

Claywork   30 minutes (depending on quantity)

Dry   2–3 days, or for immediate painting, place in front of ordinary light bulb or in warm oven 30 minutes.

Paint with underglaze colors. You may fire to 05 cone first, but this is not necessary if pieces are easily handled.

Fire 05 cone

Glaze

Fire 06 cone

MATERIALS

clay
scissors
comb
powder

sponge
soft drink bottle caps and tops, all sizes
beads
ceiling tile

brushes
teaspoon
underglazes
clear glaze

PROCEDURE

STEP 1   *Platter*   Coat hands with powder to prevent sticking. (Powder must be washed off clay before painting.) Take piece of clay the size of small grape. Coat with powder and press in your hand until it is $1/8$ inch thick. This is best done by putting clay in palm of

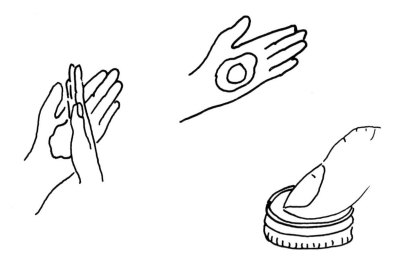

your left hand and hitting it sharply with your right hand as shown. You may cut platters by pressing the 1/8 inch thick piece of clay into a teaspoon. With the latter, be sure to smooth clay off the rim of the spoon so that when clay dries and shrinks from the spoon, it will not catch on spoon rim and break. Cut plates and platters in different sizes for variety. Smooth edges with fingers. Dry for one day, or for quicker drying, place in front of ordinary light bulb. After plates are dry, paint them with underglaze paints blue, pink, light green, or yellow *before* you place food on them.

STEP 2 Make food for plates separately, forming with fingers. Working the clay between your fingers and practicing with the items will make a finished piece easier. Allow food to dry for an hour in a warm oven or in front of a light before painting and paint the piece completely before placing on plates. (Remember that unpainted clay will be white after first firing.)

See drawings. Try to keep all food approximately 1 inch to a foot in scale. This is the generally accepted dollhouse size.

BREAKFAST

eggs—white with yellow center
sausage—brown
bread—brown
  (part of loaf may be sliced)

butter—yellow
  (part of stick may be sliced)

LUNCH OR DINNER

chicken leg—brown
boiled potato w/parsley—white
  flecked with green
carrots—orange
tomatoes—red
ham—brown edge, red center
fish—light brown, with yellow
  lemon slices

steak—red with light brown bone
turkey—brown
corn—yellow
  (use fine comb to mark kernels)
pork chops—brown
peas—green
mashed potatoes—white

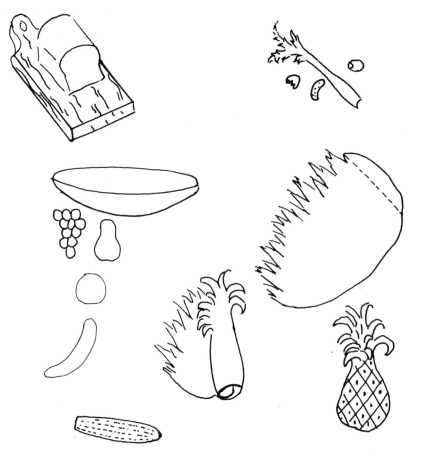

cut a one-inch-long condiment dish on which can be placed:
celery—white and pale green   pickles—white and pale green
olives—pale green, red center   radishes—red

For sandwiches on a plate, first make a loaf of bread by shaping with fingers. Let set for twenty minutes, paint outside edge brown, then slice with thin wire. Make yellow cheese and red sandwich meat to put between slices.

fruit bowl: Make flat 1½ inch plate and shape into bowl. (You may press a plastic doily into the clay to give the bowl texture and design.) Then fill with fruit:

pear—yellow   apple—red
orange—orange   grapes—purple
grapes—green   * pineapple—brown with green top
banana—yellow   peach—yellow

   * To make pineapple: cut flat piece of clay to ragged-edge shape shown above, then roll as shown and score with pointed tool.

Pixie Dresser Set

(See color plate.)

Claywork 30 minutes for each pixie
Dry 5–7 days
Fire 05 cone
Underglaze

Fire 05 cone
Glaze
Fire 06 cone

MATERIALS

clay
rolling tool
cutting tool

underglaze paints and clear glaze
or
colored glazes

sponge
slip
glaze brushes

PROCEDURE

STEP 1 Make body ¾ inch thick as shown. Bottom should be flat and top should be rounded. Push brush handle three quarters of way up through body from bottom and make three holes. This will allow air to escape during firing and prevent cracking.

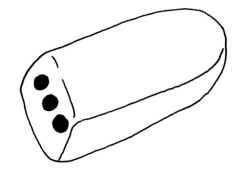

STEP 2 For legs, make coil ½ inch in diameter, 6 inches long. Form two shoes the size of small grape, as shown. Point and curl up toes as shown. Score shoe, coat with slip, and attach to each end of leg coil.

STEP 3 Score and coat middle of leg coil with slip and attach to body as shown. Cross legs into sitting position and press body firmly so pixie will balance properly. You may lay pixie on back and put legs in any position.

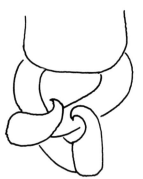

STEP 4 Make clay ball the size of pea for the neck. Score and attach to body.

STEP 5 Shape clay the size of large grape for head. Hollow head with finger. Push hole in head with brush handle, pushing from top down. The cap will cover hole.

STEP 6 To make buttons: Make four small balls the size of tiny peas. Flatten, coat with slip, and attach to front of body. With handle of brush, put small hole in middle of each button.

STEP 7 Cut cap from pattern A or B, ⅛ inch thick clay. Form into pointed cap and carefully seal seam with slip. With pointed cutting tool, make one small hole in top of cap to allow air to escape during firing.

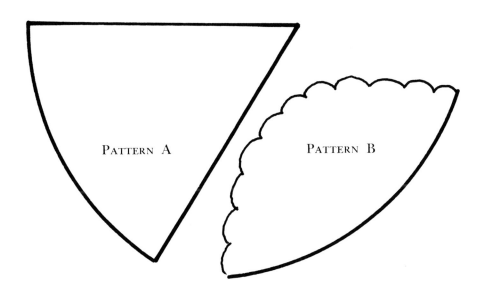

STEP 8 Coat head with slip. Place cap on head and bend over as shown.

STEP 9 Make two arms, 2½ inches long, ¼ inch in diameter. Pinch end for hand. Attach one arm to each side of body and shape as shown. Smooth ends of arms onto body with your fingers. Smooth with damp sponge.

STEP 10 Coat face with flesh underglaze. After it is dry and before firing, add face details of eyes, nose, and mouth with black under-

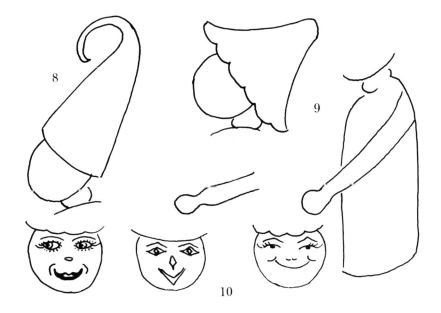

glaze using 00 pointer brush. You may glaze entire pixie in one color, or as shown in orange and green. Do not glaze face. You may use pixies as ornaments or to decorate a powder box and dresser tray as shown. Or, use them on bookends as detailed on page 88.

STEP 11   Dry five to seven days, or two days' normal drying and six to eight hours in oven.
STEP 12   Fire to 05 cone.
STEP 13   Underglaze.
STEP 14   Fire to 06 cone.
STEP 15   Clear glaze.
STEP 16   Fire to 06 cone.
  OR,   Fire to 05 cone.
        Underglaze face and hands as instructed above.
        Colored glaze on body and hat.
        Fire to 06 cone.
        (This eliminates one firing.)

PROCEDURE FOR MAKING TRAY

STEP 1   Cut tray from pattern C. The dotted line is the center of tray as the pattern forms only half of the tray.

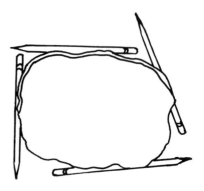

STEP 2   Smooth edges with very wet sponge. Lay tray flat on ceiling tile or board. Carefully turn up edges and put pencils around edges to hold them up until tray is dry.
STEP 3   Dry five to seven days or two days' normal drying and six to eight hours in low oven.
Fire to 05 cone.

STEP 4   Glaze with speckled yellow or light green. Place glazed pixie on side of tray and fire in this position. The pixie will become permanently attached to tray during firing when glaze melts.

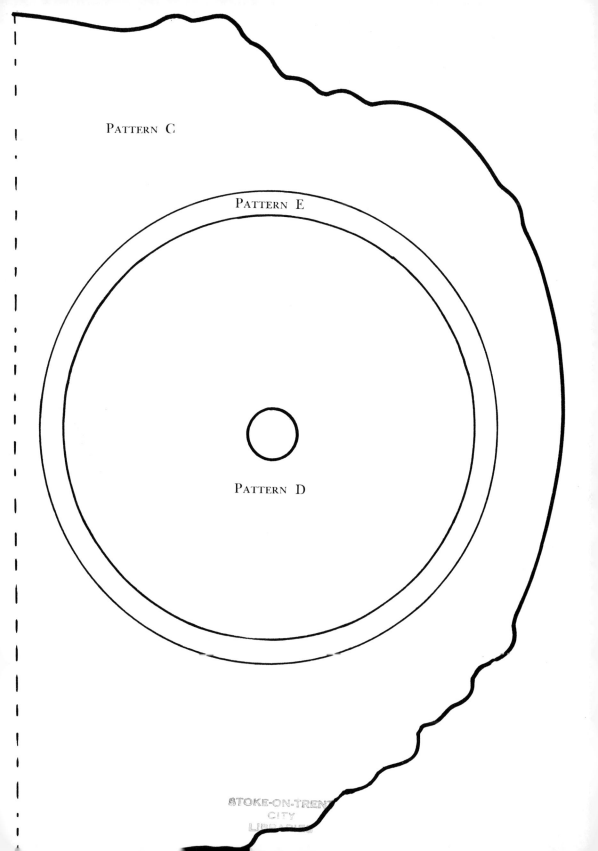

PROCEDURE FOR MAKING POWDER BOX

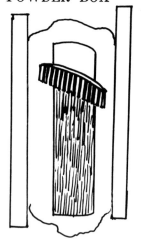

STEP 1 Roll clay ¼ inch thick. Cut a round piece of clay from pattern D for powder box base. Clean edges and place on ceiling tile or board.

STEP 2 Roll clay ¼ inch thick and 13 inches long. Cut box sides from clay. The sides should be measured with a ruler and cut 3 inches by 12¾ inches. Clean the cut edges and texture with a comb while flat.

STEP 3 Shape the box sides around outside of bottom. You should be able to do this with your fingers, sealing edges together with slip.

If you need a form inside box on which to shape the sides, wrap a large tumbler with paper towels. Set it on base, then shape sides around it—fill with crumpled paper to hold shape while drying.

STEP 4 Cut lid from pattern E. Texture upper side with comb, turning it around as you work.

STEP 5 Turn lid over. Make coil ¼ inch by 6 inches long. Score on circle on lid. Coat with slip and set coil in place. This will hold lid in place when finished.

STEP 6 Dry box five to seven days, or two days' natural drying and a warm oven six to eight hours.

STEP 7 Fire to 05 cone. During bisque firing set lid on box to prevent warping.

STEP 8 Glaze outside of box brown.
Glaze inside of box to match tray.
Set pixie, glazed and ready to fire, on lid. Firing process will attach pixie permanently to lid.

STEP 9 Fire to 06 cone. Stilt box and lid separately during glaze firing.

Grape Candle-holder

(See color plate.)

Claywork 1 hour
Dry 5–7 days

Fire 05 cone
Glaze
Fire 06 cone

MATERIALS

clay
rolling tool
cutting tool
brush

2 wooden slats
grape leaves, (real, plastic, or use pattern)
sponge

ceiling tile
underglaze paints
clear matt glaze

PROCEDURE

STEP 1 *Base* Cut round of clay 3 inches in diameter. Roll out three arms, each 2½ inches long. Score clay and attach firmly as shown. Build candle-holder on board for easy moving.

STEP 2 With piece of clay, form candle-holder in middle of clay by setting candle in and rotating. Allow ⅛ inch extra diameter for shrinkage.

STEP 3 Form thirty oval balls the size of small grapes, as shown, and place on wet towel to keep damp until needed.

STEP 4 Take ball of clay the size of grapefruit and roll out between slats. Using real, plastic, or pattern A (page 104) leaves, press in and cut out three leaves. Smooth edges of leaves with fingers and mark in veins of leaves with tool if necessary.

STEP 5 Score and coat area between arms with slip and attach first leaf.

STEP 6 Score first arm and coat with slip. Place first clump of grapes on arm and on top of first leaf edge. Where grapes are on top of each other, coat the underside grapes with slip.

STEP 7 Score and coat next area between arms with slip. Place second grape leaf. The edge toward grapes should come up and over grapes. Grapes should lay on leaf on one side.

STEP 8 Score second arm and coat with slip. Place second clump of grapes. Where grapes overlap, coat the underside grapes with slip. Grapes should rest on arm and edge of leaf.

Score and coat third area between arms and put third leaf in place.

STEP 9 Score and coat third arm for grape clump. Grapes should rest over third leaf edge and on arm. Gently raise edge of first leaf and let it lean against grapes. Do not put third grape clump on top of first leaf edge.

STEP 11 Clean all rough edges with damp sponge, and smooth grapes.

STEP 12 Put aside to dry five to seven days, or two days' natural drying and a warm oven six to eight hours.

STEP 13 You may fire to 05 cone before painting with underglaze or carefully paint the dry clay piece and fire to 05 cone.

STEP 14 Paint grapes with underglaze paints—yellow, purple, green, or pale yellow. Dab yellow brown in crevice on yellow grapes for shadow. On purple grapes, add a little black to the purple and dab into crevice. Make leaves dark green.

STEP 15 Fire to 05 cone.

STEP 16 Apply clear matt glaze.

STEP 17 Fire to 06 cone.

ALTERNATE PAINTING SUGGESTIONS:

Fire to 05 cone.

Paint leaves with green glaze.

Paint grapes with colored glaze. Variegated grape mixtures are available from most manufacturers.

When painting on grapes a variegated glaze that contains crystals, care should be taken not to have crystals near the edge where leaves meet grapes. Crystals will melt and run into leaf area.

Where glaze colors meet, use small brush and carefully connect glazes, but put only one coat of glaze for first $1/8$ inch where glazes meet.

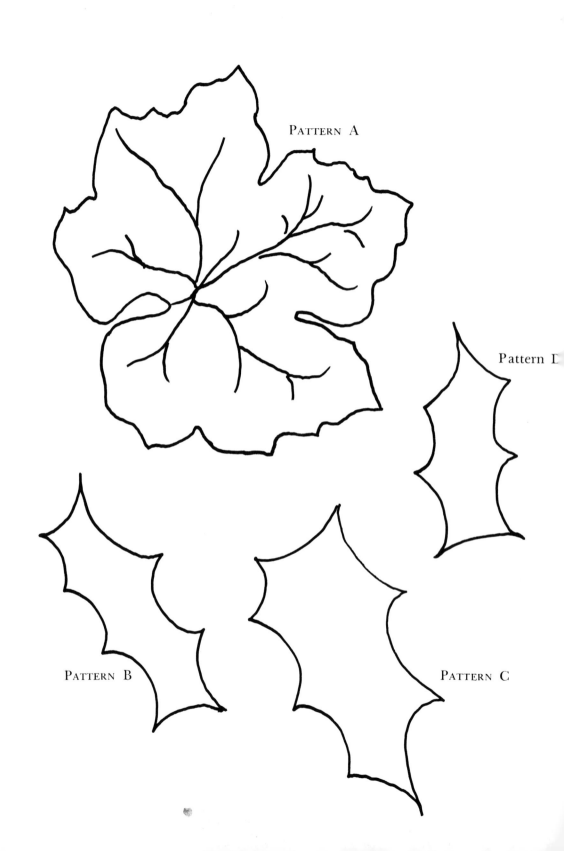

Holly Candle-holder

(See color plate.)

Claywork 1 hour
Dry 5–7 days

Fire 05 cone
Glaze
Fire 06 cone

MATERIALS

clay
rolling tool
2 wooden slats

cutting tool
green and red glaze
brush
slip

candle, 2½ inches in diameter
holly leaf *or* patterns B, C, D

PROCEDURE

STEP 1 *Base* Roll out between slats a ball of clay the size of small apple. Cut round of clay 5 inches in diameter.

STEP 2 Set candle in middle of round of clay. Make a rope of clay about 10 inches long. Score around candle, coat with slip and place rope around candle, allowing ¼ inch for shrinkage.

STEP 3 Cut twenty-five holly leaves, using patterns B, C, and D. Place on wet towel to prevent drying if you are working slowly. Cut in veins lightly with cutting tool.

STEP 4 Make twenty-five small balls the size of peas for holly berries.

STEP 5 Score and coat ring with slip. Place leaves around ring in interesting pattern. Place some leaves up on rope to cover it. Do not press tightly to candle. Allow room for shrinkage.

STEP 6 Make clumps of berries, three, five, and seven berries in a clump. Stick clumps together with slip. Do not attach to candle-holder yet.

STEP 7 Dry five to seven days, or two days' natural drying and a warm oven six to eight hours.

STEP 8 Fire to 05 cone. Fire berry clumps separately.

STEP 9 Glaze leaves with three coats of green glaze. Glaze berries separately with two or three coats of red. Place berries on leaves. Glaze will attach them in firing process. If you do not have red glaze, or if it does not fire properly, use red fingernail polish on berries after holder is finished and glazed. This polish cannot be fired and will eventually wear off, so use only when necessary.

STEP 10 Fire to 06 cone.

Rose Candle-holder

(See color plate.)

Claywork 1 hour
Dry 5–7 days

Fire 05 cone
Glaze
Fire 06 cone

MATERIALS

clay
plastic, paper pattern, or real rose leaves

cutting tool
rolling tool
kiln shelf to model on

tissue
wire screen

PROCEDURE

STEP 1 Cut round base piece 3 inches in diameter or, as shown, roll two coils 8 inches long. Scratch with cutting tool, bend for handle, and attach to 4 inch long piece of clay as shown. Place base on kiln shelf so that when finished, candleholder may be put into the kiln without disturbing.

STEP 2 Roll small, egg-shaped piece of clay for bud. Form two petals and fold petals around bud center. Place at end of base.

STEP 3 Cut pattern A for bud calyx. Place calyx over bud base. Fold back tips with fingers.

STEP 4 Roll clay 1/8 inch thick. Cut five leaves from pattern B. Or use real rose leaf. When pressed into clay, it will form veins.

STEP 5 Thin edges of leaves by pressing between fingers and curl edges up around leaf.

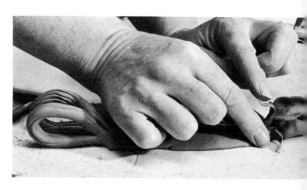

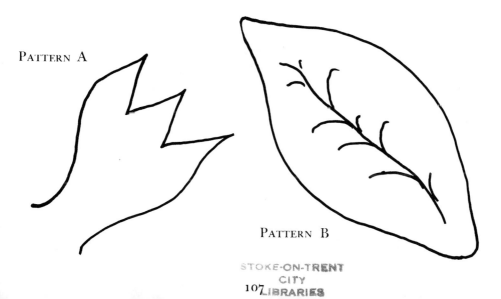

PATTERN A

PATTERN B

STEP 6   Score and coat base with slip and attach end of leaves to base as shown.

STEP 7   To form rose. Make five balls of clay the size of large cherry. Place on damp towel to keep moist.

STEP 8   Flatten edges paper thin, as shown. Taper area within dotted line to 1 inch thickness at base. Do not allow edges to split. The finished petal should be the same size as shown. Work quickly to keep clay from drying.

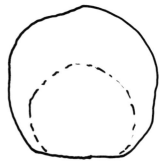

STEP 9   Cup petal in palm of hand and roll back top edge.

STEP 10   Side view of petal when ready (if cut).

STEP 11   Score outer edge of base and coat with slip. Press first petal into place.

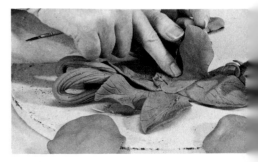

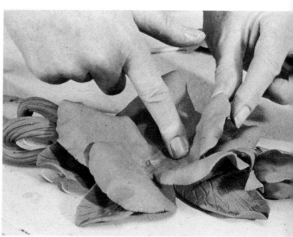

STEP 12 Outside row should take five petals overlapped like a pinwheel. When putting fifth petal in place, carefully lift edge of first petal with brush handle and slip fifth petal under edge.

STEP 13 Push down base of each petal firmly with brush handle. The thick petal base will make it easy to handle petals.

STEP 14 Form second row of petals in the same manner as in forming outside row. Set them inside outer ring. Second ring should be slightly smaller sized petals and will need three or four petals. Make petals quickly. Score and coat base with slip and press petals in with brush handle. Place petals so they cover joining of petals on outer edge. Where possible, petals should overlap. Hold petals in place with small wads of tissue.

STEP 15 There should be a space in the middle about 3/4 inch round for center of rose. Force piece of clay through 1 inch long piece of screen or strainer. Carefully lift with brush handle and place in center. With brush handle, push center stiples to side of center.

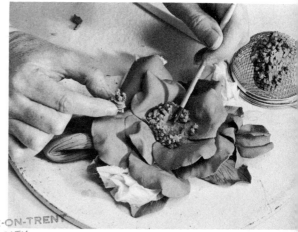

STEP 16 Using the base of a candle, carefully push down into center to form area to hold candle.

STEP 17 Dry five to seven days on kiln shelf or on half shelf. Do not move flower from shelf.

STEP 18 Place shelf into kiln and fire to 05 cone.

STEP 19 Glaze—centers yellow, petals pink, leaves and stem green.

STEP 20 Fire to 06 cone.

If making a rose to decorate a box, do not attach handle and bud. Since you will not need a center to hold a candle, you may need a third row of petals inside the rose.

Coil Items

Many interesting items can be created from coils. Roll out coils of clay between your hands or on a flat surface. Do not allow coils to become so dry that they crack.

Make napkin rings. Dry. Fire to 05 cone. Glaze with colored glazes. Fire to 06 cone.

Make bells over forms with coil handles. Smooth over bell with damp fingers to erase all joints. Seal inside with slip. Support handle with paper wad if necessary.

Make vases and bowls. Form on a round slab base. Build up sides. Score and seal with slip. Texture outside surface with cutting tool. If piece needs support while drying, fill with paper.

Dry pieces. Fire to 05 cone. Glaze and fire to 06 cone.

Tips for the Teacher

1. Be sure beginners are covered with large aprons or smocks.
2. When rolling clay, remind students to turn clay over and move it so the bottom side will be as smooth as the top side, and it will roll out round instead of long and narrow.
3. Do not let students use too much powder. Make sure it is washed off before firing. Glaze applied over excess powder will slide and leave bare spots.
4. Save clay scraps in a plastic bag. They may be wedged and reused. Wash off excess powder. If you do not have a wedging board, clay can be cut with a piece of wire and thrown on the floor onto newspaper. This is a good way to keep a restless beginner busy.
5. Give each beginner a ceiling tile to build the work on. These tiles can be easily moved to drying rack. Tiles holding flat pieces such as trivets and Christmas ornaments can be stacked and weighted so the clay pieces will not warp while drying.
6. Do not fire any piece until it is thoroughly dry. After two days' normal drying, you may hurry the drying by putting into a warm oven six to eight hours; or you can make a drying oven with a large box lined with aluminum foil with an ultraviolet lamp bulb at one end.

7. Keep tools and brushes in mugs or cans in the middle of table and have each worker put the tools back when work is finished.
8. Label tops of glazes and underglazes so they can be stored but easily spotted. A black wax pencil is good for this, or paper labels can be used.
9. Do not allow brushes to sit in glaze jar, slip pot, or water for any length of time. They must be washed and stored upside down.
10. Keep a box for patterns. They can be reused if cut from lightweight cardboard.
11. When applying underglazes, put a separate brush in each color to cut out extra washing of brushes and also eliminate possibility of a dirty brush distorting a color. Always apply three coats of underglaze.
12. Beginners can apply glazes more easily with a sponge than with a brush, particularly using matt glazes or working on large surfaces.
13. Students should not handle pieces before firing unless necessary. Unfired clay is too fragile.
14. Be sure students clean work well with damp sponge while clay is soft. This is the easy time to smooth edges and remove rough spots. After drying, a sgrafitto tool can be used if necessary.
15. Remind students to cut necessary holes. They are easily forgotten. There must be holes to allow air to escape and thus prevent cracking. Thick pieces must be completely dry and contain no air pockets or the piece may explode in the kiln.

COMMON PROBLEMS

Crazing (small cracks in glaze all over finished ware) Caused by underfired bisque. Fire all bisque one cone higher. Crazed ware may be reclaimed by refiring to the proper cone. Crazing immediately after removal from the kiln is caused by not firing the ware hot enough. Refire to proper cone. Crazing in spots may be caused by not having mixed the glaze thoroughly before using or applying too thickly.

Glaze too thin This is caused by uneven glazing or a "hard spot" on the piece. A hard spot is the result of excessive rubbing with sponge when piece is being cleaned.

Warped ware Caused by not weighting flat pieces during drying, being fired too close to the elements or firing the piece in an unnatural position.

Bubbles in glaze Caused by too heavy an application of glaze, by severe underfiring. Glaze goes through two bubbling stages, one before maturity and one after. Check glaze instructions.

Crawling of glaze (glaze moves away from clay and leaves bare spots) Caused by loose clay powder in the pores of the fired bisque. Always sponge the pieces with clear water to remove all dust or powder before glazing.

Pin holes in glazed ware Can be caused by air holes in underfired bisque or by dust particles on the bisque or glaze before firing.

Poor color in colored glaze Can be caused by overfiring a particular color. This is particularly true of red glazes, which usually fire at a lower temperature and must be well vented during firing. Apply all other colors of glaze, fire, then apply red glaze or any other special color, and fire again. Faster firing of difficult glazes helps.

Streaks in underglaze Can be caused by not applying enough coats. After a piece has been painted with underglaze and fired, you may check it for streaks by holding it under running water for a few seconds. The piece will appear glossy, and streaks and thin spots will show up. The weak spots may be touched up and refired.

Poor glaze finishes Be sure the glazed ware is loaded in the bottom of the kiln, the greenware above—the water vapor from the greenware can damage some glaze finishes.

Black specks Can be caused by organic material not completely burned out in the bisque firing. This works its way to the surface during the glaze firing.

Purple spots in gold Can be caused by a very thin application of gold or by gold that has been thinned too much.

Broken lines in gold Caused by overfiring or too heavy an application.

Clay pieces breaking during firing Caused by air pockets or piece not being properly dried.

FIRING PROBLEMS

NATURE OF TROUBLE	PROBABLE CAUSE	REMEDY
Slow firing	Low voltage	Have power company check voltage at kiln with all switches on high. Re-adjust transformer if needed.
Kiln does not reach maximum temperature	Low voltage Defective switch Broken element Blown fuse	Correct low voltage as above. Locate blown fuse or tripped circuit breaker by visual inspection and replace. Inspect element for break, if none found remove switch box and inspect element connections, replace if badly oxidized. Replace switch if no other trouble is found.
Heats in some switch positions, not in all	Blown fuse, defective element, switch or element connection	Replace fuse or reset breaker. Locate defective part and replace.
Fuse blows after kiln has fired for some time	Loose fuse or loose wire in fuse box	Tighten fuse and connections, turn off main power and sand fuse socket bright if badly oxidized.
Fuse blows as soon as switch turned to new position	Short circuit Overloaded circuit Improperly connected grounded neutral	Remove switch box, locate and correct short. Check to see if other appliances are being used on kiln circuit. Check connections in building wiring installation, plug wiring if plug has been changed.
No heat in kiln	Blown fuses, cord not plugged into outlet.	Check all fuses, including main fuses. Little probability of all elements or switches failing at once.
Hot plug or outlet	One or both defective	Replace if too hot to hold, don't fire until repaired.

GLOSSARY

Automatic shutoff Mechanical control device on kiln which shuts it off at proper temperature.
Bisque Clay piece fired once but not glazed.
Carborundum Rough stone used to remove rough glaze spots from the bottom of piece.
Ceramic A hard fired clay product.
Coil Long, thin roll of clay.
Cones Triangular pieces of clay product, used to measure temperature in kiln and constructed to bend at indicated heat.
Crazing Small crackle lines which develop during the firing process on the glazed ceramic piece.
Crystals (ceramic) Small lumps of glass product placed in glaze which runs in firing process and creates unusual effects.
Cutting tool A round, pointed tool such as a hatpin.
Double end tool Cleaning tool.
Dry footing Cleaning all unfired glaze from the bottom of pieces to be fired. This eliminates need for stilting during firing.
Greenware Unfired molded clay product.
Glaze Liquid glass product resembling cream applied to bisque. Firing gives the glassy finish.
Kiln Oven or furnace used for firing clay products to the temperature necessary to harden clay and mature glaze to a gloss finish.
Kiln wash Dry powder coating mixed with water and applied with a brush to protect fire brick in kiln from glaze drips.
Mending slip Liquid clay the consistency of thick cream, made with clay, water and drop of vinegar; used to join pieces of clay.
Nichrome wire Wire that will not melt in kiln; used to hang glazed pieces in kiln during firing.
Onionskin paper Transparent tracing paper.

Rolling tool   Rolling pin, dowling, or a piece of round handle, used to roll out and flatten clay.
Lace tool   Pointed cutting tool.
Sponge   Natural sponge.
Score   To rough clay by making crisscross lines with pointed tool.
Sgraffito tool   Tool used for cleaning and decorating.
Slats   Long strips of wood ¼ inch thick and 1–2 inches wide, used as a guide for rolling clay to proper thickness.
Stilts   Triangular posts with nichrome wire metal points used to support glazed pieces during firing process to prevent glaze from melting on floor of kiln.
Underglaze colors   Colors with clay base used on greenware or bisque before glazing.
Wedging   Process of cutting and working clay to eliminate air bubbles.

INDEX

Ashtray
    Molded Over Rocks, 44–46
    Oak Leaf, 41–43

Banks, 52–56
    Clown, 52–56
Bird feeder, 65–70
Birdhouse, 49–51
Brush care, 7, 113

Candle-holders, 101–110, 64
Candy dishes, 25, 27–29
Ceiling tile, 112
China paints, 14
Christmas ornaments from cookie cutters, 30–33
Clay 1, 8
    Cleaning, 113
    Drying, 6, 114
    Preparation of, 1
    Rolling, 3, 112
    Scraps, 1, 112
    Warped, 114
Coil work, 5, 111
Cutting tool, 4, 9

Dishes, free form, 44–48
Drying oven, 112

Figurines, 71–83
    Angel, 71–74
    Three Wise Men, 75–83
Firing, 7, 10–14
    Dry footing, 7, 14

Firing *(Cont.)*
    Drying, 7, 112
    Handling pieces, 113
    Problems, 114, 115
Free form dishes, 44–48

Gold, 14
    problems, 115
Glaze, 6
    application, 112, 114
    colored, 6
    gloss, 6
    labeling, 113
    mat, 6
    sponge application, 113
    problems, 114, 115
Grape candle-holder, 101–104

Hand modeling, 5, 86–94
    Dollhouse food, 91–94
    Horse's head, 90–91
Holly candle-holder, 105

Kiln:
    automatic shut off, 10
    cones, 10, 11
    kiln purchase, 10
    pyrometric cones, 10
    shelves, 11
    stilts, 11
    time chart, 16
    wash, 11

Lusters, 14

Modeling over forms, 4

Napkin rings, 22–26

Pansy ring, 84, 85
Paperweights, 17–20
Patio lights, 57–64
  Cat, 57–60

  Owl, 61–63

Patterns
  Angel, 72
  Bird Feeder, 66
  Bookends, 89
  Candle-holders, 105, 107
  Care of, 113
  Christmas tree ornaments, 30, 31
  Clown bank, 55
  Dollhouse food, 92
  Leaf ashtray, 42
  Napkin ring, 24
  Octopus pencil-holder, 19
  Owl Patio Light, 61
  Tray, 99
  Windchime, 37, 38, 39
  Wise Men, 76, 80, 83
Pencil-holders, 17–20

Pixies, 94–100
Powder box, 100
Problems, 114–115

Rolling tool, 8
Rose candle-holder, 106–110

Scoring, 4
Slab work, 3
Slip, 4
Supply magazines, 8

Tips for the teacher, 112–113
Tray, 98–100
Trivets, 25–26

Ultraviolet lamp bulb, 112
Underglaze paints, 6
  Application of, 113
  Problems, 115

Wedging, 1–2
  Board, 2
Wind chimes, 31–41
  Coil, 40–41
  Cookie cutter, 38
  Fish, 39
  Free form, 34–36